Lecture Notes in Physics

Edited by H. Araki, Kyoto, J. Ehlers, München, K. Hepp, Zürich
R. Kippenhahn, München, D. Ruelle, Bures-sur-Yvette
H. A. Weidenmüller, Heidelberg, J. Wess, Karlsruhe and J. Zittartz, Köln
Managing Editor: W. Beiglböck

347

Ph. Blanchard   J.-P. Dias
J. Stubbe (Eds.)

New Methods and Results in Non-linear Field Equations

Proceedings of a Conference Held at
the University of Bielefeld, Fed. Rep. of Germany,
7–10 July 1987

Springer-Verlag
Berlin Heidelberg GmbH

Editors

Philippe Blanchard
Theoretische Physik und BiBoS, Universität Bielefeld
D-4800 Bielefeld 1, FRG

Joao-Paulo Dias
CMAF, 2, Av. Prof. Gama Pinto
P-1699 Lisboa, Portugal

Joachim Stubbe
Départment de Physique Théorique, Université de Genève
CH-1211 Genève 4, Switzerland

ISBN 978-3-662-14494-7   ISBN 978-3-540-46868-4 (eBook)
DOI 10.1007/978-3-540-46868-4

© Springer-Verlag Berlin Heidelberg 1989
Originally published by Springer-Verlag Berlin Heidelberg New York in 1989
Softcover reprint of the hardcover 1st edition 1989

2153/3140-543210 – Printed on acid-free paper

INTRODUCTION

In recent years the interaction between physics and mathematics has increased mainly in the domain of non-linear field equations considered from both the deterministic and from the stochastic points of view. The two projects "Mathematics + Physics", which ran at the Centre for Interdisciplinary Research (ZiF) of the University of Bielefeld during the academic years 1975/76 and 1983/84, and more recently the BiBoS research project, supported by the Volkswagen foundation and involving the universities of Bielefeld and Bochum (the "S" stands for stochastics), are good examples of this interdisciplinary exchange (see e.g. "Mathematics + Physics Vols. I, II, III, L. Streit (ed.), World Scientific 1985, 1986, 1988, and "Trends and Development in the Eighties", S. Albeverio and Ph. Blanchard (eds.), World Scientific 1985).

Within the framework of the BiBoS project we organized a research month and a small conference on "Non-linear Fields" at the University of Bielefeld in July 1987. We want now to sketch the main subjects studied there.

Generally, one deals with a non-linear system of the form

$$\frac{du}{dt} = f\,(u,t) \qquad\qquad \text{(deterministic)}$$

or

$$du = f(u,t)\,dt + g(u,t)dw \qquad \text{(stochastic)}$$

modelled on a Banach space X. The modelling space X may be finite dimensional (e.g. IR^n) or infinite dimensional (e.g. a function space like $\mathrm{L}^2(\mathrm{IR}^n)$, $\mathrm{H}^1(\mathrm{IR}^n)$ and so on). The latter case arises in many non-linear field theoretic models involving classical non-linear wave (or Klein-Gordon) equations:

$$u_{tt} - \Delta u + kou = 0;$$

non-linear Dirac equations:

$$i\gamma^\mu \partial_\mu U + koU = 0;$$

or non-linear Schrödinger equations:

$$iu_t + \Delta u + kou = 0,$$

where kou always denotes a non-linear operator acting on the field u.

Most of the contributions to this volume deal with these equations, and typical basic problems such as the existence of solutions for a given initial state or the existence and properties of particular solutions (stationary solutions or quasi-stationary solutions, i.e. non-linear bound states, asymptotic behaviour etc.) are studied. In view of the physical applications the investigation of stochastic perturbations of such equations becomes a very interesting subject. For example, classical non-linear Klein-Gordon equations can be interpreted as the zeroth-order approximation to the "true" quantum field. Indeed, quantum effects may be modelled by means of a stochastic perturbation.

Two contributions consider aspects of stochastics and non-linear systems in this context. The lecture by S. Albeverio and A. Hilbert presents results about what is known for a finite-dimensional stochastically perturbed Hamiltonian system (existence and uniqueness of solutions, asymptotic behaviour, stability), while the article of T. Hida, J. Potthoff and L. Streit offers a short introduction to infinite-dimensional stochastic calculus (white noise analysis). for which the question of application to the systems and equations listed above is still open.

The main part of this volume concerns problems concerning the deterministic equations:

The existence of solutions for given initial data (Cauchy problem). The first lecture by J. Ginibre and G. Velo summarizes their recent perturbative results on the Cauchy problem for a class of non-linear Klein-Gordon equations with so-called local non-linear self-interactions. Their main interest is in presenting several perturbative arguments leading to the existence and uniqueness of global solutions in energy space , i.e. of solutions with finite energy.

Th. Cazenave and F.B. Weissler consider the Cauchy problem for non-linear Schrödinger equations in energy space. Their method relies on an approximation procedure leading to a limit equation. This argument allows one to relax some smoothness assumptions on the non-linear term and covers most of the previously treated cases , e.g. as obtained in the pioneering work of J. Ginibre and G. Velo on non-linear Schrödinger equations (Journ. of Funct. Anal. 32 (1979)).

J. P. Dias and M. Figueira prove the existence of solutions for a special non-linear Dirac equation which is of great interest in particle physics.

The existence and stability of non-linear bound states or solitary waves. The search for non-linear bound states of non-linear wave or Schrödinger equations leads to elliptic equations of the form

$$- \Delta u = g (u)$$

While much is known about the existence of solutions for scalar functions u (see e.g. W. Strauss, Comm. Math. Physics 55 (1977), H. Berestiycki and P.L. Lions, Arch. Rat. Mech. Anal. 82 (1983)), the situation for vector fields is different. Recent results have been obtained by H. Brezis and E. Lieb (Comm. Math. Phys. 96, (1984)). In his contribution E. Brüning proposes a modification of the Brezis-Lieb strategy to obtain ground state solutions of two-dimensional field equations.

A basic physical requirement for non-linear bound states is the condition of stability. L. Vazquez reviews several concepts of stability of stationary or quasi-stationary solutions in Hamiltonian systems in the spirit of a recently developed stability theory by M. Grillakis, J. Shateeh and W. Strauss (Journ. Funct. Anal. 74 (1984)). He also reports on recent numerical results on collision phenomena which are related to soliton stability properties of non-linear bound states.

An alternative stability theory which uses typical scaling properties of non-linear classical field theories is presented by Ph. Blanchard and J. Stubbe. It gives simple conditions for the stability of bound states of non-linear Klein-Gordon and Schrödinger equations.

Global properties of solutions. A natural question arises after solving the global Cauchy problem: how do these solutions behave for large times? In their second lecture, J. Ginibre and G. Velo present results in this direction for non-linear wave equations. They show new results on time-decay properties implied by the approximate conformal invariance of non-linear wave equations.

Finally, we would like to thank all participants for their fruitful exchange of scientific ideas. The success of the meeting was due to the speakers; thanks to their efforts it was possible to keep back of the most recent developments.

We believe that the research month in Bielefeld was an exellent starting point for close co-operation between people working on different subjects, and with different techniques, in mathematical physics.

Bielefeld   Ph. Blanchard, J.-P. Dias, J. Stubbe
July 1989

CONTENTS

Some remarks on stochastically perturbed (Hamiltonian) systems

by

Sergio Albeverio[♯]   Astrid Hilbert

Fakultät für Mathematik, Ruhr-Universität, Bochum (FRG)

ABSTRACT

We shortly mention several perturbation problems of classical dynamical systems by stochastic forces. We look more closely to the case of an Hamiltonian system consisting of a particle moving in $I\!R^d$ under the action of a force derived from a potential V and an additional stochastic force. We report on a recent extension of ours with Zehnder of results by Potter, Mc Kean, Markus and Weerasinghe. Under restrictions on the growth of V at infinity or attractiveness of the force towards the origin we give existence, uniqueness and stability results for the solution of the (stochastic) equations of motion. We also give a comparison theorem with solutions of a corresponding linearised system, via a Cameron-Martin-Girsanov-Maruyama type of formula. We also discuss the asymptotic behavior of the solution for large times, as well as the existence of a σ-finite, not finite invariant measure, the Lebesgue measure in phase space.

[♯]   BiBoS-Research Centre, SFB 237
Bochum-Düsseldorf-Essen, CERFIM Locarno (CH)

1. Introduction

Stochastic perturbations of classical dynamical systems arise in various contexts. We would like to mention briefly some of them, somewhat related methodologically, and concentrate then on the presentation of some new results concerning stochastic perturbations of finite dimensional Hamiltonian systems.

Historically one can trace the origins of the theory of stochastic perturbations of dynamical systems in considerations involving random walks and Markov chains on one hand (e.g. in work by Bachelier and Markov) and statistical mechanics on the other hand (e.g. Maxwell, Boltzmann, Einstein and Gibbs). Work by A. Einstein (1905), Smoluchowski (1905), A. Einstein and E. Hopf (1910), P. Langevin (1911), N. Wiener (1923), S. Bernstein, P. Lévy, L. Ornstein-G. Uhlenbeck (1930) (cfr. e.g. [DeH-L], [Ne1]) was very influential for later developments (they were in fact' forerunners of the modern theory of stochastic differential equations). Typically in this work the time evolution of position x and velocity v of a classical system, a Newton particle, is considered. The evolution is described by equations of the type

$$\frac{dx}{dt} = v(t), \quad \frac{dv(t)}{dt} = K\left(x(t), t\right) - \gamma v(t), \tag{1.1}$$

with $K(\cdot, t)$ a given (possibly space and time dependent) vector field over the state space $I\!R^d$ of the system, the deterministic force field. $\gamma \geq 0$ is a constant (damping). The stochastically perturbed system is obtained by adding on the right hand side a stochastic term, e.g. of the white noise type, $\dfrac{d\omega(t)}{dt}$, with $\omega(t)$ a Brownian motion on $\mathbb{R}^d$, or more generally, of the form $\sigma\left(x(t),v(t),t\right)\dfrac{d\omega(t)}{dt}$ with σ a (possibly space and time dependent) $d \times n$-matrix and $\omega(t)$ a Brownian motion on $\mathbb{R}^n$:

$$\frac{dx}{dt} = v(t), \quad \frac{dv(t)}{dt} = K\left(x(t),t\right) - \gamma v(t) + \sigma\left(x(t),v(t),t\right)\frac{d\omega(t)}{dt} \quad . \tag{1.1}'$$

Mathematically the discussion of the above equations is part of the theory of stochastic differential equations of Ito's or Stratonovich's type (with known rules to commute between the two) (see e.g. [Ar], [Ga] for introductions and [Ik-Wa], [RoW] for more advanced topics).

In the study of properties of behavior of solutions for large times, which interests us most here, methods can be differentiated according to whether the damping term is present or not and whether σ is degenerate or not. For the case with damping term ($\gamma > 0$) with σ non degenerate see e.g. [Kh].

For the case of degeneration see e.g. [Ar], [Klie].

The case $\gamma = 0$ (Hamiltonian conservative systems) is the main topics to be discussed below.

Let us note however at this point that a more abstract formulation of the problem is to look at stochastic perturbations of deterministic systems of 1. order differential equations of the form

$$\frac{dy(t)}{dt} = \beta\left(y(t),t\right), \tag{1.2}$$

with $y(t)$ a $\mathbb{R}^n$-valued function and $\beta(\cdot,\cdot)$ a given (time and space dependent) vector field. The stochastic perturbations can be of the type

$$\tilde{\sigma}\left(y(t),t\right)\frac{d\tilde{\omega}(t)}{dt},$$

with $\tilde{\omega}(t)$ a standard Brownian motion in $\mathbb{R}^m$ and $\tilde{\sigma}$ a $d \times m$-matrix-valued (time and space-dependent) function. The stochastically perturbed system is thus of the form

$$\frac{dy(t)}{dt} = \beta\left(y(t),t\right) + \tilde{\sigma}\left(y(t),t\right)\frac{d\tilde{\omega}(t)}{dt}. \tag{1.2}'$$

An important case discussed in recent years in connection with quantum physics is the one where β is a gradient field ∇u, with u built from the real and imaginary parts of a function satisfying the Schrödinger equation and $\tilde{\sigma}$ is a constant (proportional to Planck's constant), in which case one can look upon (1.2)' as the equation of stochastic mechanics ([Ne 1,2], [BlCZ]) (we remark that (1.1)' is, on the other hand, not of this type). Existence and uniquenes of solutions of this equation have been discussed (see e.g. [AHKS], [Car], [Ne2], [Bl-Go]), as well as other problems like asymptotics for $t \to \pm\infty$ ([Car]), attainability of singularities of β (see above references and [AFKS], [Fu]), and questions related to the physical interpretation of the equation (e.g. [DZ], [Tr]). Let us also mention that recently the inverse problem of finding u (i.e. β) given observations outside a bounded region has also been discussed, for the equations of stochastic mechanics ([ABKS]). There has also been much work on stochastic mechanical equations for motions on Riemannian manifolds (see e.g. [Gue], [Bl-CZ], [AB/HK1] (and references therein)), as well as on the stochastic mechanical equations for the motion of a (quantum mechanical) particle in an electromagnetic field ([Mor]). We also mention a connection between stochastic mechanical models with potentials which are Fourier transforms of measures and models of filter theory [Are1].

Another important case discussed in connection with classical dynamical systems (e.g. engineering problems) is the one where the vector field β in (1.2)'is such that zero is an equilibrium point and there exists a Ljapunov function (see e.g. [Kh]). Also the discussion of the small $\tilde{\sigma}$-limit has been pursued intensively, this limit is connected in the case of the equations of stochastic mechanics with the semiclassical limit in quantum mechanics (cfr. [Ma-Fe], [A-HK1], [Re], [Jo-M-S], [A-Ar]). We also mention that explicit connection between solutions of (1.2)' and orbits of the corresponding classical dynamical system (1.2) has been studied in the case of motions on certain manifolds with symmetries (Lie groups, symmetric spaces) (see e.g. [E], [AAH], [Are2]).

We also recall that the case of equation (1.1)' or (1.2)' where β and $\tilde{\sigma}$ are linear is extensively discussed in the literature, see e.g. [H], [GQ] (and references therein) and serves as a standard reference case in many issues.

The case of infinite dimensional state space is obviously more complicated and less results are known. For the linear case see e.g., in connection with filter theory, [Ko-Lo] (and references therein) resp. in connection with quantum fields [AHK], [Rö], [Ko]. For the nonlinear case see e.g. [AHK2], [A-Rö], [A-Ku]. The stationary case is particularly well discussed in connection with infinite dimensional Dirichlet forms (see [ARö]). For a recent development concerning a four space-time dimensional model for quantum fields see [A-HK-I]. We also mention that the equations of hydrodynamics (Euler and Navier-Stokes) with stochastic perturbation i.e.

$$\frac{du}{dt} = -(u \cdot \nabla)u + \nu \Delta u + f, \quad \text{div} u = 0 \tag{1.3}$$

$\nu = 0$ resp. $\nu > 0$ being the viscosity constant, (with suitable initial and boundary conditions), f being a deterministic plus stochastic force, have also been discussed recently in particular in connection with the proof of the existence of invariant probability measures, see [A-C], [Cr], [Fuj]. Finally let us mention that whereas all above examples are of elliptic-parabolic type, also some hyperbolic stochastic (partial) differential equations can be handled, see e.g. [ARu] (and references therein).

In this paper we shall concentrate on the case of an Hamiltonian system with finite dimensional state space. We shall see that even in this simple case relatively little is known and many interesting problems are open. More concretely we study an Hamiltonian system of the form

$$\begin{aligned} dx(t) &= v(t)dt \\ dv(t) &= K\left(x(t)\right)dt + d\omega_t, \end{aligned} \tag{1.4}$$

with ω_t a Brownian motion in $\mathbb{R}^d$, started at time 0 at the origin, i.e. $\{\omega_t, t \geq 0\}$ is a Markov stochastic process, consisting of independent Gaussian distributed random variables with independent Gaussian increments $\omega(t) - \omega(s)$, $s \leq t$, with mean zero and covariance $(t - s)$.

The inital data $x(0) = x_0$, $v(0) = v_0$ are given in $\mathbb{R}^{2d}$ (e.g. independent of the point ω in the underlying probability space). We can rewrite (1.4) in the form

$$\begin{aligned} y &\equiv \begin{pmatrix} x \\ v \end{pmatrix}, \quad \beta\left(y(t)\right) \equiv \begin{pmatrix} y(t) \\ K(x(t)) \end{pmatrix}, \\ dy(t) &= \beta\left(y(t)\right)dt + \sigma d\tilde{\omega}_t \text{ with } \tilde{\omega}_t = \begin{pmatrix} b_t \\ \omega_t \end{pmatrix}, \end{aligned} \tag{1.5}$$

b_t a Brownian motion started at time 0 in 0, independent of ω_t.

Models of this type are frequently discussed in the literature mainly numerically, in connection with vibrations in mechanical systems, wave propagation and other problems (see e.g. [Li], [Kr-S], [Sch], [Ka]).

Hamiltonian systems of this type are also obviously important in celestial mechanics (see e.g. [Mo]). One of the reasons for which they have not been so well studied on a mathematical basis is the quite intricated nature of the classical motion themselves (as compared to dissipative systems). Let us mention that a number of papers have been devoted to the study of the corresponding deterministic problem (see e.g. [Di-Z] and references therein). For stochastic perturbations of other type (multiplicative ones) see e.g. [APW], [P]. Orbits of very different long time behavior cannot be separated in finite time intervals, stable and unstable behavior being mixed. In particular, the orbits are in general neither globally stable nor asymptotically stable, see e.g. [Ar], [Mo] and [Mo-Ze]. The nature of the orbits depends in particular on the dimension of the system. For example in the case of more than three degrees of freedom the phenomenon of Arnold diffusion can met one. It is interesting to find out what one can say about perturbations in case the force is stochastic, hence typically non smooth, in particular, whether the complicated behaviour of classical orbits is enhanced or whether stochasticity is so strong as to change the picture radically. Potter [Po] (see also Mc Kean [McK]) analysed the case of a 1-dimensional nonlinear oscillator perturbed by a white noise force, described by the equations (1.4). Under assumptions on the force $K(x) = -V'(x)$, $V \in C^1(\mathbb{R})$ being attracting towards the origin, i.e. $x \cdot K(x) \leq 0$, Potter proved the existence of global solutions and results about recurrence as well as the invariance of Lebesgue measure $dxdv$ under the flow given in (1.4). These results recently have been extended in [Mark-W] who studied in particular winding numbers around the origin associated with the solution process (x, v).

Existence and uniqueness results for solutions of higher dimensional second order Ito equations, as the systems of the type given in (1.1) have been called by Borchers [Bo], have been deduced by Goldstein [Go] for systems with globally Lipschitz continuous force K, and Narita [Na] in case there exists a function, decreasing along the paths analogously to a Ljapunov function in the deterministic theory.

In the first part of this paper we shall study equations of the form (1.4) in the case where x, and v run in $\mathbb{R}^d$. In Section 2 we establish existence and uniqueness results for strong solutions of the equations, under assumptions on K which are of the type $K(x) = -\nabla V(x)$ for some $V \in C^1(\mathbb{R}^d)$, with either a condition of the form K is linear or $x \cdot K(x) \leq 0$ for $|x|$ sufficiently large or $V(x)$ sufficiently increasing at infinity. Then the solution process possesses the Markov property and continuous sample paths, furthermore it depends continuously on the initial conditions.

In section 3 we compare the solutions of the nonlinear system (1.4) with the ones of a corresponding linear system given by

$$dx = vdt$$
$$dv = -\gamma xdt + dw \tag{1.6}$$

and γ a constant dxd-matrix with positive eigenvalues. This is done by establishing a Comeron-Martin-Girsanov-Maruyama type formula for the Radon Nikodym derivative of the probability measures. We apply these results to prove some properties which hold with probability one for the nonlinear system by exploiting their validity for the associated linear system.

In section 4 we recover some features of the behavior of the solution process of the nonlinear system for large times. In particular, we give estimates for the energy functional of the process. We introduce the generator of the diffusion, solving (1.4), and show its hypoellipticity (in the sense that the coefficient functions span the tangent space to phase space). By a Hörmander's type theorem one obtains the absolute continuity of the transition probability of the transition probability w.r. to Lebesgue measure without further restrictions but continuity of the coefficient functions. We also show that for the solution process of (2.1) an existence and uniqueness theorem for a σ-finite invariant measure, which is the "normalized" Lebesgue measure, holds. For more details on the result presented here we refer to [AHZ], [H].

2. Existence and Uniqueness of Solution

We consider the Hamiltonian system with stochastic force given by the stochastic differential system

$$dx = vdt$$
$$dv = K(x)dt \quad + \quad dw_t \tag{2.1}$$

where $t \in \mathbb{R}_+$ is time, $x(t)$ is position in $\mathbb{R}^d$ at time t, $v(t)$ is velocity at time t.
$K(\cdot)$ is a deterministic force, $(w_t, \mathcal{F}_t, t \in \mathbb{R}_+)$ is a Brownian motion in $\mathbb{R}^d$ started at the origin at time zero, see section 1 for motivations. The initial conditions are given.
It is useful to introduce the phase space variable $y = (x, v) \in \mathbb{R}^{2d}$ and to write (2.1) in the form

$$dy = \beta(y)dt + \sigma d\tilde{w}_t \tag{2.2}$$

with

$$\beta(y) \equiv \begin{pmatrix} v \\ K(x) \end{pmatrix}, \quad \sigma \equiv \begin{pmatrix} 0 & 0 \\ 0 & 1 \end{pmatrix}, \quad \tilde{w}_t = \begin{pmatrix} b_t \\ w_t \end{pmatrix},$$

where $(b_t, \mathcal{A}_t, t \in \mathbb{R}_+)$, $\mathcal{A}_t = \sigma\{b_s, s \leq t\}$, is an $(\mathcal{F}_t)$ independent Brownian motion in $\mathbb{R}^d$ issued from 0 at time 0. The initial condition $y(0)$ is given. For simplicity we state the theorem assuming $y(0) = 0$.

Theorem 2.1

Each of the following conditions is sufficient for the existence of pathwise solutions of (2.1), (2.2) for all $t \in \mathbb{R}_+$:
a) 1) $|K(\alpha) - K(\beta)| \leq C_1|\alpha - \beta| \quad \forall |\alpha|, |\beta| < R$, for some constants R, C_1 .
    2) $|K(a)| \leq C_2(1 + |\alpha|) \quad \forall \alpha \in \mathbb{R}^d$, for some constant C_2 .

b) 1) $\alpha \mapsto K(\alpha)$ is a locally Lipsschitz function from $\mathbb{R}^d$ into $\mathbb{R}^d$. Moreover
    2) For $d \geq 1$: $K(\alpha) = -\nabla V(\alpha)$ for some $V \in C^1(\mathbb{R}^d)$
    3)$\alpha \cdot K(\alpha) \leq 0$ for all $\alpha \in \mathbb{R}^d$.

Remark

Corresponding statements holds for $t \geq t_0$ with initial condition $y(t_0)$ given, and
$(\alpha - y(t_0)) \cdot K(\alpha) \leq 0$.

Proof:

Statement a) can be proven by a stochastic version of Picard-Lindelöf method of iteration, see e.g. [Ar1] (Cor. 6.3.4) .
1) For $d = 1$ the statement b) in a special case of a result of Potter [Po], see e.g. [McK], [Na1]. We give a proof valid for $d \geq 1$ which uses a Ljapunov function for the solution of the stochastic differential equation. To this end we introduce the energy functional:

$$W(y) = \frac{1}{2}|v|^2 + V(x) - V(0). \tag{2.3}$$

Let L be the differential operator (generator) associated with (2.2) i.e.

$$L = v \cdot \nabla_x + K(x) \cdot \nabla_v + \frac{1}{2}\Delta_v \tag{2.4}$$

where $\nabla_x (\nabla_v)$ are the gradient w.r. to $x(v)$, respectively, and Δ_v is the Laplacian w.r. to v. All operators are acting on functions of $(x, v) \in \mathbb{R}^{2d}$. Applying L to the energy function W we find

$$LW = \frac{1}{2} \qquad (2.5)$$

Furthermore from condition b)2)

$$\alpha \cdot K(\cdot) = -\alpha \cdot \nabla V(\alpha) = |\alpha| \frac{\partial V(\alpha)}{\partial |\alpha|} \qquad (2.6)$$

we conclude for $|\alpha| \neq 0$

$$V(\alpha) = V(0) - \int_0^{|\alpha|} \frac{1}{|\beta|} (\beta \cdot K(\beta)) d|\beta| \geq V(0) , \qquad (2.7)$$

where we used assumption b)3).
For the energy functional in (2.3) this implies

$$W(y) \geq \frac{1}{2}|v|^2 . \qquad (2.8)$$

and

$$|\sigma \nabla_y W(y)|^2 = |v|^2 \leq 2W(y).$$

Since β in locally Lipschitz continuous, following [Na2] and [Ik-Wa] (Def. 2.1) we can discuss local solutions.
In a first step we introduce stopping times

$$\sigma_n(w) \equiv \inf\{t \geq 0| \ |y(t)| \geq n\}$$

of the process $Y \equiv (y(t), t \geq 0)$ and define the <u>explosion time</u> $e(0)$ of Y for given innitial condition $y(0) = 0$, by

$$e(0) = \sup_{n \in N} \{\inf \ t| \ |y(t)| \geq n\} \qquad (2.9)$$

with inf replaced by $+\infty$ if the set is empty. For any $n \in \mathbb{N}$ the process $Y_n \equiv (y(t \wedge \sigma_n)| \ t \geq 0)$ defines a local solution of (2.1) on the ball $B_n(0)$, since there exists a uniform Lipschitz constant on every $B_n(0)$ which guarantees the existence and pathwise uniqueness of solutions. For $n \to \infty$ the local solutions Y_n converge a.s. to the maximal solution of (2.1), corresponding to the martingale

$$y^i(t) - y^i(0) - \left[\sum_{k=1}^d \int_0^t \sigma_k^i(y(s)) d\tilde{w}_k(s) + \int_0^t \beta^i(y(s)) ds \right] \qquad i = 1, \ldots, d$$

with expectation 0, for $t \in [0, e(0)]$ cfr. [Na3]. It is shown in [Na2] that the proof of existence of a global solution Y for initial condition $y(0) = 0$ is equivalent to an infinite explosion time $e(0)$, i.e. to

$$N_0 \equiv \left\{ e(0) < \infty, \text{ and } \lim_{t \uparrow e(0)} |y(t)| = +\infty \right\} \qquad (2.10)$$

satisfying

$$\mathbb{P}(N_0 = \{e(0) < \infty\}) = 1. \qquad (2.11)$$

For times before an explosion occurs we can reexpress the energy functional W given by (2.3) by applying Ito's formula to the differential

$$dW = v(t) dw(t) + \frac{d}{2} dt + o(t^{\frac{3}{2}})$$

obtaining

$$W(y(t)) = W(0) + \int_0^t v(s)dw(s) + d\frac{t}{2} \; . \tag{2.12}$$

The process $W(y(t))$ can be simplified by introducing analogously to [McK] a new Brownian motion

$$a(\tau(t)) \equiv \int_0^t v(s)dw_s$$

with a clock running according to the time

$$\tau(t) = \int_0^t |v(s)|^2 ds.$$

Under the assumption of Theorem 2.1 b) a global solution of (2.1) in established due to (2.11) by the following

Lemma 2.2

Adopting the hypothesis of Theorem 2.1 b) there holds

$$\mathbb{P}(e(0)) = 1 \quad .$$

Proof:

The proof of the higher dimensional statement can be reduced to the one for the one dimensional case in [Po] with y being replaced by $|y|$. The proof is by contradiction, distinguishing the cases $\tau(e(0)) < \infty$ and $\tau(e(0)) = \infty$, and using the sample path properties of the Brownian motion $a(\cdot)$. Thus the a.s. finiteness of $|y(t)|$, where $0 \le t \le e(0)$, is deduced, which yields the contradiction.

Theorem 2.3

Each of the conditions a) and b) of Theorem 2.1 is sufficient also for pathwise uniqueness of solutions of the equations (2.1), (2.2), i.e. if y, y' are two solutions of (2.2) on the same probability space with the same filtration s.t. $y(0) = y'(0)$ a.s., then

$$y(t) = y'(t) \qquad\qquad \forall t \ge 0 \text{ a.s.}$$

Proof:

a) This case is covered e.g. by [Fr].
b) This case follows from [Ik-Wa], (Theorem 3.1 p. 164) since the coefficients of the equations (2.1),(2.2) are in particular locally Lipschitz continuous. This yields uniqueness for times $t, 0 \le t \le e(0)$, and this together with the fact (demonstrated in the proof of Theorem 2.1) that $e(0) = \infty$ a.s. yields uniqueness for all $t \ge 0$.
It also follows from [Ik-Wa] that (2.1), (2.2) have unique strong solutions.

3. A Girsanov Formula

In this section we shall investigate whether the probability measure associated with the solution of the stochastic differential equation (2.1), i.e.

$$dy(t) = \beta(y(t))dt + \sigma d\tilde{w}_t$$

with

$$\beta(y) = \begin{pmatrix} v \\ K(x) \end{pmatrix}, \quad \sigma = \begin{pmatrix} 0 & 0 \\ 0 & 1 \end{pmatrix}, \quad \tilde{w} = \begin{pmatrix} b_t \\ w_t \end{pmatrix},$$

is absolutely continuous with respect to the probability measure associated with the corresponding Gaussian process given by the stochastic differential equation

$$d\eta(t) = a(\eta)dt + \sigma d\tilde{w}_t \tag{3.1}$$

with

$$\eta = \begin{pmatrix} z \\ u \end{pmatrix}, \qquad z, u \in \mathbb{R}^{2d}$$

$$a(\eta) = \begin{pmatrix} u \\ -\gamma z \end{pmatrix}, \qquad \gamma \text{ a constant matrix},$$

and vice versa. If this holds, almost sure statements concerning the nonlinear system are equivalent to almost sure statements concerning the linear system.
In order to show the equivalence we shall derive a (Cameron-Martin-Maruyama-) Girsanov formula relating the probability measures.

Lemma 3.1

Let $W(y(t)) \equiv \frac{1}{2}|v(t)|^2 + V(x(t)) - V(x_0)$ with V s.t. $-\nabla V = K$, $Y = (y(t), t \geq 0)$ satisfying (2.1) or (2.2). Then, for all $t \geq 0$

$$\text{i)} \quad \mathbb{E}(W(y(t))) = \mathbb{E}(W(0)) + d\frac{t}{2}$$

and

$$\text{ii)} \mathbb{E}(W^2(y(t))) = \mathbb{E}(W(0)^2) + \frac{d}{2} \int_0^t \mathbb{E}\left(\frac{|v(s)|^2}{2} + V(x(s)) - V(0)\right) ds$$

$$+ \frac{1}{2} \int_0^t \mathbb{E}(|v(s)|^2) ds.$$

Proof:
Statement i) follows from 2.12 by taking expectation, and using that $\int_0^t v(s) \cdot dw(s)$ is a martingale with expectation zero.
ii) Given $F \in C^2(\mathbb{R})$, and a solution $Y = (y(t), t \geq 0)$ of (2.1) or (2.2) for $W(y(t)) = \frac{1}{2}|v(t)|^2 + V(x(t)) - V(x_0)$ we calculate, using Ito formula successively,

$$F(W(y(t))) = F(W(0))) + \int_0^t F'(W(y(s)))v(s) \cdot dw(s) +$$

$$\frac{1}{2} \int_0^T \left\{ dF'(W(y(s))) + |v(s)|^2 F''(W(y(s))) \right\} ds.$$

Inserting $F(\lambda) = \lambda^2$, $\lambda \in \mathbb{R}$, we get the equation ii) of the lemma. ∎

Lemma3.2

Let Y be the solution of the stochastic differential equation (2.2). Then the configuration process $(x(t), t \geq 0)$ as well as the velocity process $(v(t), t \geq 0)$ possess finite absolute moments of second order.

Proof:

Starting from (2.8) we find using Lemma 3.1 i)

$$
\begin{aligned}
I\!\!E(|v|^2) &\leq 2I\!\!E(W(y(t))) + 2C_k \\
&= 2W(y_0) + dt + 2C_k
\end{aligned}
\tag{3.2}
$$

Inserting equations (2.2) and (3.2) into the expression for the second moment of the configuration process then there holds:

$$
\begin{aligned}
I\!\!E(|x|^2) &= I\!\!E\left(|\int_0^t v(s)ds|^2\right) \\
&\leq \int_0^t I\!\!E(|v(s)|^2)ds \\
&\leq \frac{d}{2}t^2 + 2(W(y_0) + C_k)t
\end{aligned}
\tag{3.3}
$$

where C_k is a positive constant.

Lemma3.3

Let K, $Y = ((x(t), v(t)), o \leq t \leq T)$ and $\eta = ((z(t), u(t)), u \leq t \leq T)$ be as in Theorem 2.1 and in (3.1) respectively. There holds

$$
I\!\!P\left\{\int_0^T |K(x(t)) + \gamma x|^2 ds < \infty\right\} =
$$

$$
I\!\!P\left\{\int_0^T |K(z(t)) + \gamma z(t)|^2 ds < \infty\right\} = 1
$$

Proof:

As was shown in Lemma 3.2 the second absolute moment of the configuration process $X = (x(t), t \geq 0)$ is finite (see (3.3)). Since K is Lipschitz continuous and X has continuous paths $I\!\!P$ a.e which are bounded on $[0, T]$ we have

$$
\int_0^t |K(x(t)) + \gamma x(t)|^2 dt \leq 2 \int_0^T \left(|K(x(t))|^2 + |\gamma x(t)|^2\right) dt
$$

$$
\leq 2 \int_0^T \left(K_T^2 |x(t) - x_0|^2 + |\gamma|^2 |x(t)|^2 + |K(x_o)|^2\right) dt < \infty,
\tag{3.4}
$$

where K_T, the Lipschitz constant of the sphere $B_{r_0}(x_0)$, $r_0 = \max_{t \in [0,t]}\{|x(t) - x_0|\}$, may depend on the specific path ω. Since η is a degenerate Gaussian process we find analogously $I\!\!P$ a.e

$$
\int_0^T |K(z(t)) + \gamma z(t)|^2 dt \leq 2 \int_0^T \left(|K(z(t))|^2 + |\gamma z(t)|^2\right) dt
$$

$$
\leq 2 \int_0^T \left(\bar{K}_T^2 |z(t) - z_0|^2 + |\gamma|^2 |z(t)|^2 + |K(x_o)|^2\right) dt < \infty
\tag{3.5}
$$

where $\bar{K}_T^2$ the Lipschitz constant of the sphere $B_{s_0}(z_0)$, $s_0 = \max_{t \in [0,t]}\{|z(t) - z_0|\}$, may depend on ω, and the second constant factor is defined as $|\gamma|^2 \equiv |\text{trace } \gamma|^2$. ∎

Proposition 3.4

Let $T > 0$, and $(w_t, t \in [0,T])$ be a Wiener process with state space $\mathbb{R}^n$. Furthermore take $(\Omega, \mathcal{F}, \mathbb{P})$ as Wiener space, and $\mathcal{F}_t \equiv \sigma(w_s, s \in [0,T])$ as the σ-algebra generated by the Wiener process up to time T. Let ξ_t, η_t be stochastic processes on $(\Omega, \mathcal{F}, \mathcal{F}_t, \mathbb{P})$ s.t. ξ_t and η_t are $\mathcal{F}_t$-measurable and assume ξ_t, η_t satisfy the stochastic differential equations

$$d\xi_t = A(t,\xi)dt + \sigma(t,\xi)dw_t$$
$$d\eta_t = a(t,\eta)dt + \sigma(t,\eta)dw_t,$$

with $\eta_0 = \xi_0$ being $\mathcal{F}_0$-measurable and

$$\mathbb{P}\{\xi_0 < \infty\} = 1.$$

A, a, σ should satisfy the following conditions:

i) $A(t,\xi), a(t,\eta), \sigma(t,\xi), \sigma(t,\eta)$ are $\mathcal{F}_t$-measurable and such that there exist unique strong solutions ξ_t, η_t of the above stochastic differential equations(i.e. solutions on the probability space $(\Omega, \mathcal{F}, \mathbb{P})$).

ii) For arbitrary fixed $t \in [0,T]$ the system of algebraic equations in x over $\mathbb{R}^n$ admits a solution $\alpha(t,x)$:

$$\sigma(t,x)\alpha(t,x) = A(t,x) - a(t,x).$$

The function $\alpha(\cdot, \cdot)$ should be measurable in $t \in [0,T]$, $x \in \mathbb{R}^n$, and satisfy

iii)

$$\mathbb{E}\exp\left(\frac{1}{2}\int_0^T |\alpha(t,\xi_t)|^2\,dt\right) < \infty \quad .$$

Let μ_ξ resp. μ_η be the probability measures on $(\Omega, , \mathcal{F}, \mathbb{P})$ associated with ξ resp. η in the sense that the probability of $\{\xi_{t_1} \in B_1, \ldots, \xi_{t_n} \in B_n\}$ is $\mu_\xi(\xi_{t_1} \in B_1, \ldots, \xi_{t_n} \in B_n)$ for all $B_i \in \mathcal{B}(\mathbb{R}^n)$, and all $t_i \in [0,T]$, and similarly for μ_η Then μ_ξ and μ_η are mutually absolutely continuous i.e. are equivalent and one has $\mathbb{P}$-almost surely

$$\frac{d\mu_\eta}{d\mu_\xi} = e^{-\int_0^T \alpha(t,\xi)\cdot dw_t} \cdot e^{-\frac{1}{2}\int_0^T |\alpha(t,\xi)|^2\,dt}.$$

Proof:

The proof is a multidimensional version of the statement made in [Lip] (Th. 5.4 p. 160) for $\mathbb{R}^1$. ∎

Theorem 3.5

Give the probability space $(\Omega, \mathcal{A} \otimes \mathcal{F}, \mathbb{P})$ as above. Let $Y = ((x(t), v(t)), t \geq 0)$ with initial data y_0 be the global solution of the nonlinear stochastic differential equation (2.2) with K satisfying the assumptions of the existence and uniqueness Theorem 2.1.

Let $\eta = ((z(t), u(t)), t \geq 0)$ be the solution of the linear stochastic differential equation (3.1). Then the process η is equivalent to Y, in the sense that the probability measures μ_η and μ_y constructed on path space Ω are equivalent. The Radon-Nikodym derivatives are given by

$$\frac{d\mu_y}{d\mu_\eta}(\eta) = \exp\left(+\int_0^T (K(z(t)) + \gamma z(t)) \cdot dw(t) - \frac{1}{2}\int_0^T |K(z(t)) + \gamma z(t)|^2\,dt\right).$$

and

$$\frac{d\mu_\eta}{d\mu_y}(y) = \exp\left(-\int_0^T (K(x(t)) + \gamma x(t)) \cdot dw(t) - \frac{1}{2}\int_0^T |K(x(t)) + \gamma x(t)|^2 dt\right).$$

Proof:

For arbitrary fixed $t \in [0, T]$, and $a(y)$, $\beta(y)$, σ as in Lemma 3.2 the system of algebraic equations in x over $\mathbb{R}^{2d}$ admits a measurable solution $\bar{\alpha}(x)$

$$\sigma\bar{\alpha}(y) = \beta(y) - a(y) \quad . \tag{3.6}$$

Denote by $\bar{\alpha}_1, \bar{\alpha}_2$ the first respectively last d components of $\bar{\alpha}(x)$. $\bar{\alpha}_1$ is left undefined by (3.6), for convenience we choose $\bar{\alpha}_1(y) = 0$. $\bar{\alpha}_2$ is determined by (3.6) as

$$\bar{\alpha}_2(y) \equiv \alpha(x) \equiv K(x) + \gamma x \quad . \tag{3.7}$$

For any function $\kappa = (\kappa_1, \kappa_2)$, $\kappa_i \in C([0, T], \mathbb{R}^d)$, $i = 1, 2$, we define the exit time of $h(t, x) := \int_0^t |\alpha(x(s))|^2 ds$ in (3.5) from the sphere of radius n by

$$\vartheta_n(\kappa_1) := \begin{cases} \inf\{t \mid t \le T\}, & h(t, \kappa_1) \ge n \\ T & \text{otherwise} \end{cases} \tag{3.8}$$

and set $\Theta_n(\kappa) := \vartheta_n(\kappa_1)$. The characteristic function $\chi_n(t, \kappa) = \chi_{\{\Theta_n(\kappa) \ge t\}}$ allows to define a truncated drift coefficient A_n relating by a Girsanov transformation the solution of (3.12) below with (3.1). A_n is defined by (3.9,3.10):

$$K_n(\kappa_1) = -(\gamma\kappa_1) + \chi_n(t, \kappa)[K(\kappa_1) + \gamma\kappa_1] , \tag{3.9}$$

$$A_n(\kappa) = \begin{pmatrix} \kappa_2 \\ K_n(\kappa_1) \end{pmatrix} . \tag{3.10}$$

Let us consider the process $Y^{(n)}(t) = (y_t^{(n)}, \mathcal{A}_t \otimes \mathcal{F}_t | 0 \le t \le T)$ defined by

$$y_t^{(n)} := y_{t \wedge \Theta_n(y)} + \int_0^t [1 - \chi_n(s, Y^{(n)})](-\gamma y^{(n)}(s))ds + \int_0^t [1 - \chi_n(s, Y)^{(n)}]\sigma \, dw_s . \tag{3.11}$$

By Proposition (2.8) equation (3.11) has a unique solution, with $y^{(n)}(t) = y(t)$ for $t \le e(0)$. Applying Ito's calculus we find that $Y^{(n)}$ satisfies the following stochastic differential equation

$$dy^{(n)}(t) = A_n(y^n(t))dt + \sigma dw_t , \quad Y_{n,0} = y_0 \quad . \tag{3.12}$$

Using

$$A_n(\kappa) - \begin{pmatrix} \kappa_2 \\ -\gamma\kappa_1 \end{pmatrix} = \chi_n(t, \kappa_1)[K(\kappa_1) + \gamma\kappa_1]$$

we deduce

$$\int_0^t |K_n(x^{(n)})) + \gamma x^{(n)}|^2 \le n \quad \mathbb{P} \text{ a.e.} \quad .$$

Moreover, the criterion for the Girsanov density to be a martingale given in Proposition (3.4) is fulfilled, i.e we have

$$\mathbb{E} \exp\left[-\int_0^T \left(K_n(x^{(n)}) + \gamma(x^{(n)})\right) \cdot dw(t) - \frac{1}{2}\int_0^T |K_n(x^{(n)}(s)) + \gamma(x^{(n)}(s))|^2 ds\right] = 1, \tag{3.13}$$

and the following formula for the density holds:

$$\frac{d\mu_{Y(n)}}{d\mu_\eta}(\eta) = \exp\left[\int_0^T (K_n(\eta(s)) + \gamma\eta(s)) \cdot dw(s) \cdot dw(s) - \frac{1}{2}\int_0^T |K_n(\eta(s)) + \gamma\eta(s)|^2 ds\right]$$

$$= \exp\left[\int_0^{T\wedge\Theta_n(\eta)} \alpha(\eta(s)) \cdot dw(s) - \frac{1}{2}\int_0^{T\wedge\Theta_n(\eta)} |\alpha(\eta(s))|^2 ds\right] \tag{3.14}$$

$$\equiv \zeta_{T\wedge\Theta_n(\eta)}(\eta).$$

Let Γ be a Baire set in the space $C([0,T], \mathbb{R}^{2d})$ Then we approximate $\mu_Y(\Gamma)$ by the density of the truncated process and rewrite the expression gained in this way by using (3.14)

$$\mu_Y(\Gamma) = \lim_{n\to\infty} \mu_{Y(n)}(\Gamma \cap \{\Theta_n(\kappa) = T\})$$

$$= \lim_{n\to\infty} \int_{\Gamma\cap\{\Theta_n(\kappa)=T\}} \zeta_{T\wedge\Theta_n(\kappa)}(\kappa)\, d\mu_\eta(\kappa)$$

$$= \lim_{n\to\infty} \int_{\Gamma\cap\{\Theta_n(\kappa)=T\}} \zeta_T(\kappa)\, d\mu_\eta(\kappa)$$

$$= \int_\Gamma \zeta_T(\kappa)\, d\mu_\eta(\kappa)$$

This means that μ_ξ is absolutely continuous with respect to μ_η, and the Radon-Nikodym derivative is given by ζ_T.

Finally, let us point out that (3.5) also implies that the stochastic integral $\int_0^T (K(z(t)) + \gamma z(t)) \cdot dw(t)$ exists and is finite ($\mathbb{P}$ a.e), see e.g. [McKean], 2.3.6 (p. 25) or [Lip], Note 7 (pp. 104). Then

$$\mu_\eta\{\zeta_T(\kappa) = 0 \mid \kappa \in \operatorname{supp}\mu_\eta\} = 0, \tag{3.15}$$

which ends the proof. ∎

Remark 3.6

All assumptions are satisfied e.g. in the following cases:

a) for some constants $\alpha_0, C_1, C_2 > 0$:

$$d = 1, \qquad V(\alpha) = C_1\alpha^{2q} + C_2$$

$$d > 1, \qquad V(\alpha) = C_1|\alpha|^{2q} + C_2, \quad q \in \mathbb{N}, \ \forall|\alpha|^2 > \alpha_0$$

b) for $q \in \mathbb{N}$

$$d = 1, \quad V(\alpha) = \alpha^{2q}f(\alpha) \text{ with } f \text{ smooth and such that}$$

$$\alpha f'(\alpha) + 2qf(\alpha) \geq 0 \quad \forall|\alpha| \geq \alpha_0,$$

$$V(\alpha) \text{ Lipschitz for } |\alpha| \leq |\alpha_0|, \text{ for some } \alpha_0 > 0.$$

Remark 3.7

The restriction to initial condition $y_0 = (0,0)$ is technical, and can be released since in the autonomous case all we need is a growth condition far away from the origin, see [H]. Following [Mark-W] from the Girsanov-type stochastic equivalence result we can transfer some conclusions about the behaviour of the linear process in finite time to the nonlinear system. In particular we have:

Under the assumptions of Theorem 3.5

$$x(t)^2 + v(t)^2 > 0 \qquad \forall t > 0,$$

provided

$$x(0)^2 + v(0)^2 > 0 \qquad \text{almost surely..}$$

4. Some Additional Remarks

We can use the energy function defined in (2.15) to obtain some estimates on the behaviour of the solution process Y of equation (2.2) with initial condition $y_0 = 0$ as $t \to \infty$. In fact we have

Theorem 4.1

Under the assumptions of Theorem 2.1 the process $W(y(t)) - d\frac{t}{2}$, $0 \leq t < \infty$, is a martingale, and we have

$$\left[W(0) + d\frac{t}{2} \right]^n \leq E\left([W(y(t))]^n \right) \leq d^n \sum_{k=1}^{n} \prod_{l=1}^{k} (2l - 1) \binom{2n}{2k} [W(0)]^{n-k} (d\frac{t}{2})^k .$$

Proof:

As was pointed out in section 2 see (2.12), $W(y(t)) - W(0)$ is the sum of a martingale and the nonrandom function $d\frac{t}{2}$.

a) Applying Ito's formula to $(W(y(t)))^n$, repeatedly, and using that the martingale $\int_0^t v(s)(W(y(s)))^{n-1} \cdot dw_s$ has expectation zero, we find taking expectation

$$E(W(y(t))^n) - E(W(y_0)^n) =$$
$$= \int_0^t \left(d\frac{n}{2} E\left(W(y(s))^{n-1} \right) + \frac{n(n-1)}{2} E\left(|v|^2 W(y(s))^{n-2} \right) \right) ds , \quad n \geq 2. \qquad (4.1)$$

From (4.1) and

$$0 \leq E\left(|v(s)|^2 W(y(s))^{n-2} \right) \leq 2E\left(W(y(s))^{n-1} \right)$$

we have

$$d\frac{n}{2} \int_0^t E\left(W(y(s))^{n-1} \right) ds \leq E(W(y(t))^n) - E(W(0)^n) \qquad (4.2)$$

$$\leq n(n - 1 + \frac{d}{2}) \int_0^t E\left(W(y(s))^{n-1} \right) ds. \qquad (4.3)$$

Moreover, one can easily see by induction that

$$E(W(y(t))^n) \leq d^n E\left((\sqrt{t}dz + \sqrt{W(0)})^{2n} \right), \qquad (4.4)$$

where the expectation on the r.h.s. is with respect to standard Gaussian measure z (with mean 0 and covariance $\frac{1}{2}$). Computing the expectation on the r.h. side in (4.4) yields the right estimate given in this theorem. The proof of the left inequality goes by induction. We integrate the inequality given by the assumption of the induction, and insert the resulting estimate into (4.2). ■

The process of Theorem 4.1 is a Markov diffusion process, since it solves the stochastic equation (2.2). The Markov kernel $P(t, a, db)$, $a, b \in \mathbb{R}^{2d}$, defined by the transition probability is then well defined. Since $K(x)$ is continuous by our assumptions, $P(t, a, db)$ defines a (Feller) Markov semigroup on $C_b(\mathbb{R}^{2d})$. Let L be its infinitesimal generator s.t. for $f \in C_0^\infty(\mathbb{R}^{2d})$

$$(Lf)(a) = \frac{d}{dt}\mathbb{E}^a(f(y(t)))|_{t=0}.$$

Using Ito's formula, see e.g. [Fr], [Si]:

$$(Lf)(a) = (\Delta_v + K(x) \cdot \nabla_v + v \cdot \nabla_x) f(a) \tag{4.5}$$

with $a \equiv (x, v)$. Following [Po] one can show that $P(t, a, db)$ is absolutely continuous w.r. to Lebesgue-measure db for fixed t, a. This is seen by looking at the transition probability kernels $P_n(t, a, db)$ for the approximation of (2.2) obtained by replacing $K(x)$ by

$$K_n(x) = \begin{cases} K(x) & \text{if } |x| \leq n \\ K(n) & \text{if } |x| > n. \end{cases}$$

By known results on the fundamental solution of degenerate parabolic equations with globally Lipshitz coefficients we have that $P^n(t, a, db) = p^n(t, a, b)db$, with $p^n(t, a, b) \in$
$\in L^1_{loc}(db)$. A dominated convergence argument shows that $P^n(t, a, A) \to P(t, a, A)$ and from $P^n(t, a, A) = 0$ for $|A| = 0$ follows that $P(t, a, A) = 0$, hence the absolute continuity of $P(t, a, db)$.

Let us regard $P(t, a, db)$ as defining a Markov semigroup T_t in the space $\mathcal{M}$ of signed measures with finite total variation, by defining for $\mu \in \mathcal{M}$

$$T_t\mu(\cdot) \equiv \int_{\mathbb{R}^{2d}} P(t, a, \cdot)\mu(da). \tag{4.6}$$

We call μ an <u>invariant measure for the Markov semigroup</u> T_t, or an <u>invariant measure</u> for the process Y if

$$T_t\mu(A) = \mu(A) \tag{4.7}$$

for any Borel subset A of $\mathbb{R}^{2d}$ and all $t \geq 0$. We shall see that under the above assumptions P has a density p w.r. to Lebesgue measure. We have for (4.4) by Fubini's Theorem

$$\mu(A) = \int_{\mathbb{R}^{2d}} P(t, a, A)\mu(da) = \int_A \left(\int_{\mathbb{R}^{2d}} p(t, a, y)\mu(da) \right) dy \tag{4.8}$$

where we used the absolute continuity of $P(t, a, \cdot)$ w.r. to Lebesgue measure. From (4.8) absolute continuity of the invariant measure follows.
At this point, we insert the following

Remark 4.2

Let $\tilde{L}$ be the infinitesimal generator of the semigroup T_t, then any invariant measure μ of the process Y of Theorem 2.1 satisfies

$$\tilde{L}\mu(db) = 0, \tag{4.9}$$

and conversely.

Lemma 4.3

The Lebesgue measure on $\mathbb{R}^{2d}$ is an invariant measure for Y.

Proof:

Let $\tilde{L}^*$ be the formal dual of the operator $\tilde{L}$ with $\mathcal{D}(L) \subset C_b(\mathbb{R}^{2d})$. For P_t the dual of T_t given by:

$$\int_{\mathbb{R}^{2d}} (P_t f)(b) \mu(db) = \int_{\mathbb{R}^{2d}} f(b) T_t \mu(db) \qquad \forall f \in C_0^2(\mathbb{R}^{2d})$$

we get by differentiating w.r. to t:

$$\int_{\mathbb{R}^{2d}} f(b) \tilde{L} \mu(db) = \int_{\mathbb{R}^{2d}} L f(b) \mu(db),$$

hence, on $C_0^2(\mathbb{R}^{2d})$, there holds

$$L = \left(\tilde{L} \right)^* .$$

This implies using the special form (4.5) of L:

$$\tilde{L} = \Delta_v - K(x) \cdot \nabla_v - v \cdot \nabla_x \quad . \tag{4.14}$$

In particular, applied to Lebesgue measure λ this yields

$$\tilde{L} \lambda = 0 \quad . \tag{4.11}$$

By Remark 4.2 λ is then an invariant measure of the process Y solving (2.1) (or (2.2)). ∎

Remark 4.4

One verifies easily that, if K is C^∞, L is hypoelliptic in the sense that it has the form

$$L = X_1^2 + X_0 \tag{4.12}$$

with $X_{i_1} = \frac{\partial}{\partial v_i}$, $X_{i_0} = K_i(x) \cdot \frac{\partial}{\partial v_i} + v_i \cdot \frac{\partial}{\partial x_i}$, $1 \le i \le d$, so that $\{X_{i_1}, [X_{i_1}, X_{i_0}]| \ 1 \le i \le d\}$ span the smooth vector fields over $\mathbb{R}^{2d}$.
For additional results see [AHZ], [H].

Obviously many open problems remain, to name only one of them: is the solution process null recurrent (cfr. [Bat])?

ACKNOWLEDGEMENTS

This paper is based on joint work with Edy Zehnder [AHZ]. We are very grateful to him for the joy of collaboration and to Professors Phillippe Blanchard, Antonella Calzolari, Eric Carlen, Gianfausto Dell' Antonio, Detlev Dürr, David Elworthy, Shigeo Kusuoka, Wilfried Loges, Gianna Nappo, Leonid Pastur, Ludwig Streit for very interesting discussions.

REFERENCES

[A] S. Albeverio, *Some new developments concerning Dirichlet forms, Markov fields and quantum fields*, SFB 237 Preprint, to appear in Proc IAMP Conf. Swansea, Edts. J. Davis, A. Truman

[AAH] S. Albeverio, T. Arede, Z. Haba, *On the left invariant Brownian motions and heat kernels of nilpotent Lie groups*, SFB 237-Preprint (1988)

[AAr]   S. Albeverio, T. Arede, *The relation between quantum mechanics and classical mechanics: a survey of some mathematical aspects*, pp37-76 in **Chaotic Behaviour in Quantum Systems, Theory and Applications**, ed. G. Casati, Plenum Press, NY(1985)

[ABHK1] S. Albeverio, Ph. Blanchard, R. Høegh-Krohn, *Diffusions sur une variété riemannienne, barrières infranchissables et applications*, Astérisque (Colloque en l'Honneur de Laurent Schwartz) **132**, 181-202 (1985)

[ABHK2] S. Albeverio, Ph. Blanchard, R. Høegh-Krohn, *Feynman path integrals and the trace formula for Schrödinger operators*, Comm. Math. Phys. **83**, 49-76 (1982)

[ABKS] S. Albeverio, Ph. Blanchard, S. Kusuoka, L. Streit, *An inverse problem for stochastic differential equations*, BiBoS-Preprint (1989)

[AC]   S. Albeverio, A.B. Cruzeiro, *Global flows with invariant (Gibbs) measures for Euler and Navier-Stokes two dimensional fluids*, BiBoS-Preprint, to appear Comm. Math. Phys. (1989)

[AFKS] S. Albeverio, M. Fukushima, W. Karwowski, L. Streit, *Capacity and quantum mechanical tunneling*, Comm. Math. Phys. **81**, 501-513 (1981)

[AGQ]   S. Albeverio, Guanglu Gong, Minping Quian, in preparation

[A-HK1] S. Albeverio, R. Høegh-Krohn, *Oscillatory integrals and the method of stationary phase in infinitely many dimensions, with application to the classical limit of quantum mechanics I*, Inv. Math **40**, 59-106(1977)

[A-HK2] S. Albeverio, R. Høegh-Krohn, *Diffusion fields, quantum fields and fields with values in Lie groups*, pp. 1-98 in Ed. M. A. Pinsky, **Stochastic Analysis and Applications**, M. Dekker, New York (1984)

[AHKI] S. Albeverio, R. Høegh-Krohn, K. Iwata, *Covariant Markovian random fields in four space-time dimensions with nonlinear electromagnetic interaction*, pp. 69-83 in Edts. P. Exner, P. Šeba, Lect. Notes Phys. **324** Springer (1989)

[AHKS] S. Albeverio, R. Høegh-Krohn, L. Streit, *Energy forms, Hamiltonians and distorted brownian paths*, J. Math. Phys. **18**, 907-917 (1977)

[AHZ]   S. Albeverio, A. Hilbert, E. Zehnder, *Hamiltonian systems with a stochastic force: nonlinear versus linear, and a Girsasnov formula*, in preparation

[Al-Ku] S. Albeverio, S. Kusuoka, *Maximality of infinite dimensional Dirichlet forms and Høegh-Krohn's model of quantum fields*, to appear in R. Høegh-Krohn Memorial Volume

[Am]   H. Amann, **Gewöhnliche Differentialgleichungen**, W. de Gruyter 1983

[APW]   L. Arnold, G. Papanicolaou, V. Wihstutz, *Asymptotic analysis of the Lyapunov exponent and rotation number of the random oscillator and applications*, SIAM J. Appl. Math. **46**, 427-450 (1916)

[Ar]   L. Arnold, **Stochastic Differential Equations**, J. Wiley & Sons 1971

[Are1] T. Arede, *A class of solvable nonlinear filters*, Stochastics **23**, 377-389 (1988)

[Are2] T. Arede, Lisboa Thesis, in prep.

[Ar-K] L. Arnold, W. Kliemann, **On Unique Ergodicity for Degenerate Diffusions**, Bremen Report **147** (1986), Forschungsschwerpunkt Dynamische Systeme, Bremen

[ARö1] S. Albeverio, M. Röckner, *Classical Dirichlet forms on topological vector spaces - the construction of the associated diffusion process*, BiBoS Preprint, to appear in Prob. Theory rel. Fields (1989)

[ARö2] S. Albeverio, M. Röckner, Proc. Ascona 88, in preparation

[ARu]   S. Albeverio, F. Russo, in preparation

[Bat]   R.N. Battacharya, S. Ramasubramanian, *Recurrence and Ergodicity of Diffusions*, J. Mult. Anal. **12**, 95-122 (1982)

[Bl-C-Z] Ph. Blanchard, Ph. Combe, W. Zheng, *Mathematical and Physical Aspects of Stochastic Mechanics*, Lect. Notes Phys **281**, Springer, Berlin (1987)

[Bl-Go] Ph. Blanchard, S. Golin, *Diffusion Processes with Singular Drift fields*, Comm. Math. Phys. **109**, 421-435(1987)

[Bor] D.R. Borchers, *Second Order Stochastic Differential Equations and related Ito Processes*, Ph.D. Thesis, Carnegie Institute of Technology 1964

[Car] E. Carlen, *Existence and sample path properties of the diffusions in Nelson's stochastic mechanics*, Stochastic Processes - Mathematics and Physics (Edts. S. Albeverio, Ph. Blanchard, L. Streit), Lecture Notes Maths. **1158**, Springer, Berlin (1986)

[Cr] A.B. Cruzeiro, *Solution et mesures invariantes pour des équations d'évolution stochastiques du type Navier-Stokes*, Exp. Math. **7**, 73-82 (1989)

[DeH-L] G. L. De Haas-Lorentz, *Die Brownsche Bewegung und einige verwandte Erscheinungen*, Vieweg, Braunschweig (1913)

[Die] R. Dieckerhoff & E. Zehnder, *An A-Priori Estimate for Nonlinear Oscillatory Differential Equations*, Ann. Scuola Norm. Pisa **14**, 79-95 (1987)

[Do] C. Doléans-Dade, *Quelques applications de la formule de changement de variables pour les semimartingales*, Zeitschr. Wahrsch. verw. Geb. **16**, 181-194 (1970)

[DZ] D. Dürr, N. Zanghi, in preparation, to appear in Proc. 2nd Intern. Conference, Ascona 1988, Edts. S. Albeverio, G. Casati, U. Cattaneo, D. Merlini, R. Moresi, World Scient. (1989)

[E] D. Elworthy, to appear in Proc. Ref. [DZ]

[Fr] A. Friedman, **Stochastic Differential Equations and Applications**, Vol. 1, Academic Press, New York 1975

[Fu] M. Fukushima, *Energy forms and diffusion processes*, Mathematics and Physics, Lectures on Recent Results, Vol 1, (Ed. L. Streit), World Scientific (1985), pp. 65-97

[Fuj] H. Fujita-Yashima, Pisa Preprint, in preparation

[Ga] Th. C. Gard,*Introduction to Stochastic differential equations*, M. Dekker, New York (1988)

[GQ] Guanglu Gong, Minping Quian, *Singular perturbation, winding number and symmetry of drifted Brownian motions*, SFB-237 Preprint (1988)

[Go] J.A. Goldstein, *Second Order Ito Processes*, Nagoya Math. J. **36** (1969), 27-63

[Gue] F. Guerra, *Quantum field theory and probability theory. Outlook on new possible developments,* in: S. Albeverio, Ph. Blanchard (Edts.), *Trends and Developments in the Eighties"*, World Sient.,Singapore (1985)

[H] A. Hilbert, Ph.D., Bielefeld, in preparation

[Ik-Wa] N. Ikeda, S. Watanabe, **Stochastic Differential Equations and Diffusion Processes**, North-Holland, Amsterdam 1981

[JoMS] G.Jona-Lasinio,F.Martinelli,E.Scoppola, *Multiple tunnelings in d-dimensions of a quantum particle in a hierarchical potential*, Ann. I. H. Poincaré **42**, 73 (1985)

[Ka] N.G.V.Kampen, **Stochastic Processes in Physics and Chemistry**, North Holland, Amsterdam(1981)

[Kh] R.Z. Khas'minskiï, **Stochastic Stability of Differential Equations**, Sijthoo, Alphen aan den Rijn(1980)

[Kli] W.Kliemann, *Recurrence and Invariant measures for Degenerate Diffusions*, Ann. Prob. **15**, 690-707(1987)

[Kol] T. Kolsrud, *Gaussian random fields, infinite dimensional Ornstein-Uhlenbeck processes, and symmetric Markov processes*, Acta Appl. Math.

[Ko-Lo] T. Koski, W. Loges, *Asymptotic statistical interference for a stochastic heat flow problem*, Statistics and Probability Letters **3**, 185-189 (1985)

[Kr-S] P.Krée,C.Soize, **Mécanique Aléatoire**, Dunod(1983)

[Ku] H.J. Kushner, **Approximation and Weak Convergence Methods for Random Processes, with Applications to Stochastic Systems Theory**, MIT Press, Cambridge 1984

[Li] A.J. Lichtenberg, M.A. Lieberman, **Regular and Stochastic Motion**, Springer Verlag (Applied Mathematical Sciences **38**) 1983

[Lip] R.S. Lipster, A.N. Shirjaev, **Statistics of Random Processes I: General Theory**, Springer Verlag 1977

[Ma-Fe] V.P.Maslov,M.V.Fedoriouk,*Semi-Classical Approximation in Quantum Mechanics*, D.Reidel, Dordrecht(1981)

[McK] H.P. McKean, **Stochastic Integrals**, Academic Press New York 1969

[Mark-W] L. Markus, A. Weerasinghe, *Stochastic Oscillators*, J. Diff. Equ. **21**, 288-314(1988)

[Mo] J. Moser *Stable and Random Motion in Dynamical Systems, with Special Emphasis on Celestial Mechanics* in: Ann. Math. Studies **77**, Princeton U.P., Princeton N.J. 1973

[Mor] L.M. Morato, *On the dynamics of diffusions and the related general electromagnectic potentials*, J. Math. Phys. **23**, 1020-1024 (1986)

[Mo-Ze] J. Moser, E. Zehnder, book in preparation

[Na1] K. Narita, *No explosion criteria for stochastic differential equations*, J. Math. Soc. Japan **34** 192-203 (1982)

[Na2] K. Narita, *Explosion time of second-order Ito processes*, J. Math. Anal. Appl. **104**, (1984)418-427

[Na3] K. Narita, *On explosion and growth order of inhomogeneous diffusion processes*, Yokohama Math. J. **28**, 45-57 (1980)

[Ne1] E. Nelson, *Dynamical Theories of Brownian Motion*, Princeton University Press, Princeton (1967)

[Ne2] E. Nelson, *Quantum Fluctuations*, Princeton University Press, Princeton (1985)

[P] M. A. Pinsky, *Instability of the harmonic oscillator under small noise*, SIAM J. Appl. Math. **46**, 451-463 (1980)

[Po] J. Potter, *Some Statistical Properties of the Motion of a Nonlinear Oscillator Driven by White Noise*, Ph.D. Thesis, M.I.T. (1962)

[Re] J. Rezende, *The method of stationary phase for oscillatory integrals on Hilbert spaces*, Comm. Math. Phys.

[Rö] M. Röckner, *Traces of harmonic functions and a new path space for the free quantum field*, J. Funct. Anal.

[RoW] G. Rogers, D. Williams, Diffusions, Markov processes and martingales, Vol. 2, J. Wiley, Chichester (1987)

[Sch] Z. Schuss,*Theory and application of stochastic differential equations*, J. Wiley, New York (1980)

[Si] B. Simon, **Functional Integration and Quantum Physics**, Academic Press, New York 1979

[St-V] D.W. Stroock, S.R. Varadhan, **Multidimensional Diffusion Processes** Springer Berlin 1979

[Tr] A. Truman, J. T. Lewis, *The stochastic mechanics of the ground-state of the hydrogen atom*, pp. 168-179 in "**Stochastic Processes - Mathematics and Physics**", Edts. S. Albeverio, Ph. Blanchard, L. Streit, Lect. Notes Maths. **1158**, Springer, Berlin (1986)

[Ze] E. Zehnder, *Some perspectives in Hamiltonian systems*, in "**Trends and Developments in the Eighties**", Edts. S. Albeverio, Ph. Blanchard, World Scient., Singapore, 1985

STABILITY OF GROUND STATES

FOR NONLINEAR CLASSICAL FIELD THEORIES

Ph. Blanchard and J. Stubbe

Theoretische Physik and BiBoS
Universität Bielefeld
D-4800 Bielefeld 1

0. Introduction

In this contribution we want to describe a stability theory for solitary waves
of Hamiltonian systems which arise from many models in classical field theories.
Most classical field theoretic models have common properties, e.g. scaling behaviour,
existence of a ground state etc. [4]. Our abstract version will contain these ingredients. This will enable us to extend previous results by the authors [2,3] and
to present an alternative approach to the theory of Grillakis, Shatah and Strauss
[7]. To illustrate the general context let us start with an well known example
namely the nonlinear logarithmic Schrödinger equation. On $\mathbb{R}^n$ $n \geq 3$ we consider
the equation

(NLS) $\qquad i\phi_t + \Delta\phi + f(\phi) = 0$

where $f(\phi) = \phi \log |\phi|^2$. The associated energy is given by

(0.1) $\qquad E(\phi) = K(\phi) - V(\phi)$

with

(0.2a) $\qquad K(\phi) = \frac{1}{2} \int_{\mathbb{R}^N} |\nabla\phi|^2 \, dx$

(0.2b) $\qquad V(\phi) = \frac{1}{2} \int_{\mathbb{R}^N} |\phi|^2 \, (-1 + \log |\phi|^2) \, dx$.

$E(\phi)$ is, at least formally, conserved for solutions of (NLS). $E(\phi)$ is well-defined in the class

(0.3) $\qquad W \equiv H^1(\mathbb{R}^n) \cap \{\phi \in L^1_{loc} \mid |\phi|^2 \log |\phi|^2 \in L^1\}$.

It was proven by Cazenave [5] that W is a reflexive Banach space (so-called
Sobolev-Orlicz space) and that E is of class C^1 on W. See also [13].

For the moment we restrict ourselves to the subspace of radial functions in
W, i.e. we consider only functions in

(0.4) $\qquad X \equiv W_r = \{\phi \in W , \quad \phi \text{ radial}\}$.

We can write equation (NLS) also in the standard Hamiltonian form

(H) $\qquad \dfrac{d\phi}{dt} = J \, E'(\phi) \qquad$ on X

where J is simply the multiplication by $(-i)$. This Hamiltonian system is invariant under gauge transformations

(0.5) $\qquad U(s)\phi = e^{is} \phi \quad , \quad s \in \mathbb{R}$.

The invariance implies the conservation of the charge

(0.6) $\qquad Q(\phi) = -\dfrac{1}{2} \int_{\mathbb{R}^n} |\phi|^2 dx$

which is well-defined on X .

We are interested in solitary wave solutions of (H) of the form

(0.7) $\qquad \phi(t,x) = U(\omega t) \, \phi_\omega(x) \qquad , \quad \omega \text{ fixed.}$

They satisfy the "stationary" equation

(0.8) $\qquad - \Delta \, \phi_\omega = \phi_\omega \log |\phi_\omega|^2 - \omega \, \phi_\omega$

which can be equivalently written in the form $E'(\phi) = \omega \, Q'(\phi)$.

The existence of nontrivial solutions of (0.8) can be proved by variational methods as follows:

Associated to the stationary problem we define the "action" functional

(0.9) $\qquad L_\omega(\phi) \equiv E(\phi) - \omega \, Q(\phi)$

$\qquad\qquad\qquad = K(\phi) - V_\omega(\phi)$

where $V_\omega = V + \omega Q$. In particular we are interested in solutions ϕ_ω which have least action among all possible nontrivial solutions of the stationary equation. Such solutions are called ground states.

To obtain ground state solutions one solves the following constrained minimization problem (see e.g. [1,9]):

(0.10) $\qquad I(\omega) = \inf \{K(\phi), \phi \in X , V_\omega(\phi) = 1\}$.

The fact that a solution of this problem can be transformed into a solution of the field equation relies essentially on the nice behaviour of K and V_ω under the action of a scaling group T_σ .

Indeed, let

(0.11) $T_\sigma \phi(x) = \phi(x/\sigma)$

i.e. T_σ is a representation of the group of dilations, then we have

(0.12a) $K(T_\sigma \phi) = \sigma^{n-2} K(\phi)$

$V_\omega(T_\sigma \phi) = \sigma^n V_\omega(\phi)$.

This property of K and V_ω will be called "scale covariance".

In the case of the logarithmic Schrödinger equation the ground state is explicitly known and given by

(0.13a) $\phi_\omega(x) = \phi_o \exp(-\omega/2)$

and

(0.13b) $\phi_o(x) = \exp(N/2 - 1/2\ x^2)$.

Now we are interested in the Liapunov stability of the ϕ_ω-orbit $\{U(s)\phi_\omega, s \in \mathbb{R}\}$ in X .

For the logarithmic Schrödinger equation the result is well-known: The first proof of orbital stability was given by Cazenave [5] using compactness methods. Recently the same result was obtained by Blanchard, Stubbe and Vazquez [2] who extended the methods of two earlier papers by Shatah and Strauss [11,12] .

The most general approach to the stability problem has been given very recently by Grillakis, Shatah and Strauss [7] who studied the stability of solitary waves of Hamiltonian systems in a real Banach space. The main tool of their theory is the linearized Hamiltonian

(0.14) $H_\omega \equiv E''(\phi_\omega) - \omega\ Q''(\phi_\omega)$.

The idea is to show that the Liapunov stability of the ϕ_ω-orbit is equivalent to the fact that ϕ_ω minimizes locally the energy E subject to constant charge ϕ . Moreover these properties are equivalent to the condition that the action of the solitary wave $L_\omega(\phi_\omega)$ considered as a function of ω is convex. Unfortunately this theory is not applicable to the logarithmic Schrödinger equation since the energy functional is not twice differentiable as pointed out in [2] and thus the linearized operator H_ω does not exist. Our main goal is to present a stability theory for ground state solitary waves of Hamiltonian systems for which the linearized Hamiltonian does not exist.

In Section 1 we describe our abstract framework and state the main results. In Section 2 we give the outline's of the proofs. Technical details are omitted. They will be published elsewhere. In Section 3 we present some examples and in Section 4 we discuss some extensions, in particular to systems invariant under more than one symmetry.

1. The abstract model

Let X be a real Banach space. We consider the following Hamiltonian system on X

$$\text{(H)} \qquad \frac{du}{dt} = J\, E'(u)$$

where $J : X^* \xrightarrow{\text{onto}} X$ is a skew-symmetric linear operator from X^* onto X and $E : X \longrightarrow \mathbb{R}$ is a C^1 functional on X .

We are now in a position to list our main **assumptions**:

1.1 Properties of the energy   E

E can be splitted into two scale covariant parts K and V i.e. there exists a continuous mapping

$$\text{(1.1)} \qquad \sigma \longmapsto T_\sigma \in L(X,X) \quad , \quad \sigma \in \mathbb{R}^+$$

such that

$$\text{(1.2)} \qquad T_{\sigma_1}\, T_{\sigma_2} = T_{\sigma_1 \sigma_2} \quad , \quad T_1 = \text{Id}_X$$

and there exist C^1 functionals K and V such that

$$\text{(1.3)} \qquad E(u) = K(u) - V(u)$$

and

$$\text{(1.4a)} \qquad K(T_\sigma u) = \sigma^r\, K(u)$$

$$\text{(1.4b)} \qquad V(T_\sigma u) = \sigma^s\, V(u)$$

with $0 < r < s$, i.e. $\gamma \equiv r/s \in (0,1)$.

K and V (and hence E) are invariant under a strongly continuous one parameter group of isometries $U(s) : X \dashrightarrow X$, $U(s+r) = U(s)\, U(r)$, i.e.

$$\text{(1.5a)} \qquad K(U(s)u) = K(u)$$

$$\text{(1.5b)} \qquad V(U(s)u) = V(u) \qquad .$$

Associated to this invariance there is another conserved quantity

(1.6) $Q(u) = \frac{1}{2} <Bu,u>$

where $B : X \to X^*$ is symmetric and JB is an extension of the infinitesimal generator $U'(0)$. $<,>$ denotes the dual pairing between X and X^*.

1.2 Property of the 'charge' Q

We assume that Q has the following behaviour under the action of the scaling group T_σ.

(1.7) $Q(T_\sigma u) = \sigma^q Q(u)$ with $q = r$ or $q = s$

We define

(1.8a) $K_\omega(u) \equiv K(u) - \omega\delta_{rq} Q(u)$

(1.8b) $V_\omega(u) \equiv V(u) + \omega\delta_{qs} Q(u)$

and assume $K_\omega(u) > 0$ if $u \neq 0$ and $K_\omega(0) = V_\omega(0) = 0$.

Furthermore we define

(1.9) $L_\omega(u) \equiv E(u) - \omega Q(u)$
$= K_\omega(u) - V_\omega(u)$.

1.3 The minimization principle

We assume that there exist $\omega_1 < \omega_2$ such that for any $\omega \in (\omega_1,\omega_2)$ the problem

(1.10) $I(\omega) = \inf \{K_\omega(u) \mid u \in X , V_\omega(u) = 1\}$

has a unique solution $\widetilde{\phi}_\omega$, i.e.

$K_\omega(\widetilde{\phi}_\omega) = I(\omega)$, $V_\omega(\widetilde{\phi}_\omega) = 1$

in the sense that for any minimizing sequence $(u_n)_n$ (i.e. $V(u_n) = 1$ and $\lim_{n\to\infty} K_\omega(u_n) = I(\omega)$) there exists a subsequence (u_{n_j}) and a sequence of real numbers (s_{n_j}) such that

(1.11) $U(s_{n_j})u_{n_j} \longrightarrow \widetilde{\phi}_\omega$ in X as $n_j \longrightarrow \infty$.

This means that any minimization sequence for (1.10) tends to the orbit $\{U(s)\widetilde{\phi}_\omega, s \in \mathbb{R}\}$.

If we define $\phi_\omega \equiv T_{\tilde\sigma(\omega)} \tilde\phi_\omega$ with $\tilde\sigma(\omega) = (\gamma I(\omega))^{1/s}$ then ϕ_ω is a ground state of the stationary equation

(1.12) $\qquad L'_\omega(\phi_\omega) = 0$

and $u(t) = U(\omega t)\phi_\omega$ is a solitary wave solution of (H) .

A consequence of the scaling property is the following virial theorem for ϕ_ω .

(1.13) $\qquad \gamma K_\omega(\phi_\omega) = V_\omega(\phi_\omega)$.

We define

(1.14) $\qquad d(\omega) \equiv L_\omega(\phi_\omega)$

then $d(\omega)$ satisfies $d(\omega) = (1 - \gamma)K_\omega(\phi_\omega) > 0$ by (1.13) and $d'(\omega) = - Q(\phi_\omega)$.

Furhtermore we have the following alternative characterizations of $d(\omega)$:

(1.15a) $\qquad d(\omega) = \inf \{L_\omega(v) \mid v \neq 0 , \; \gamma K_\omega(v) - V_\omega(v) \leq 0\}$

(1.15b) $\qquad d(\omega) = \inf \{L_\omega(v) \mid K_\omega(v) = K_\omega(\phi_\omega)\}$.

<u>Definition:</u> The ϕ_ω-orbit $\{U(\omega t)\phi_\omega, t \in \mathbb{R}\}$ is stable if for all $\varepsilon > 0$ there exists a $\delta > 0$ such that if $\|u_0 - \phi_\omega\| < \delta$ and $u(t)$ is a solution of (H) with $u(o) = u_0$ existing for any $t \geq 0$ and

$$\sup_{t>0} \; \inf_{s \in \mathbb{R}} \; \|u(t) - U(s)\phi_\omega\| < \varepsilon .$$

Otherwise we call the ϕ_ω-orbit unstable.

Before describing our results it will be convenient first to consider a one-dimensional mechanical analogue of our abstract model (see also [10]).

Qualitatively $L_\omega(\omega)$ has the same form as the energy for the one-dimensional motion in a potential well. In our context L_ω turns out to be the energy of the 'modulated' Hamiltonian system

(H_{mod}) $\qquad \dfrac{dv}{dt} = J L'_\omega(v)$

which we obtain from (H) by the transformation $u(t) = U(\omega t)v(t)$ where $u(t)$ is a solution of (H) .

For arbitrary fixed nonzero $u \in X$ we consider the function

$$l(\sigma) = L_\omega(T_\sigma u)$$
$$= \sigma^r K_\omega(u) - \sigma^s V_\omega(u) .$$

If $V_\omega(u) \leq 0$ then $l(\sigma)$ is increasing and goes to infinity as $\sigma \to \infty$.

If $V_\omega(u) > 0$ then $l(\sigma)$ as a unique maximum $\sigma^* = \sigma^*(u) = (\gamma K_\omega(u)/V_\omega(u))^{1/s-r}$ and $l(\sigma) \to \infty$ as $\sigma \to \infty$.

We calculate the height

$$(1.16) \qquad h_\omega(u) \equiv L_\omega(T_{\sigma^*}u) = (1 - \gamma)\sigma^{*r} K_\omega(u) > 0 .$$

We now define the lowest height to be passed by

$$d(\omega) = \inf_{u \neq 0} h_\omega(u) .$$

Equivalently, if we always normalize u so that $\sigma^* = 1$, i.e. $\gamma K_\omega(u) - V_\omega(u) = 0$, we recover the variational characterization (1.15a).

Hence ϕ_ω lies on the 'mountain pass' of the energy mountains when travelling from zero to regions far away. The height of the mountain pass is exactly $d(\omega)$. This situation is sketched in the figure below.

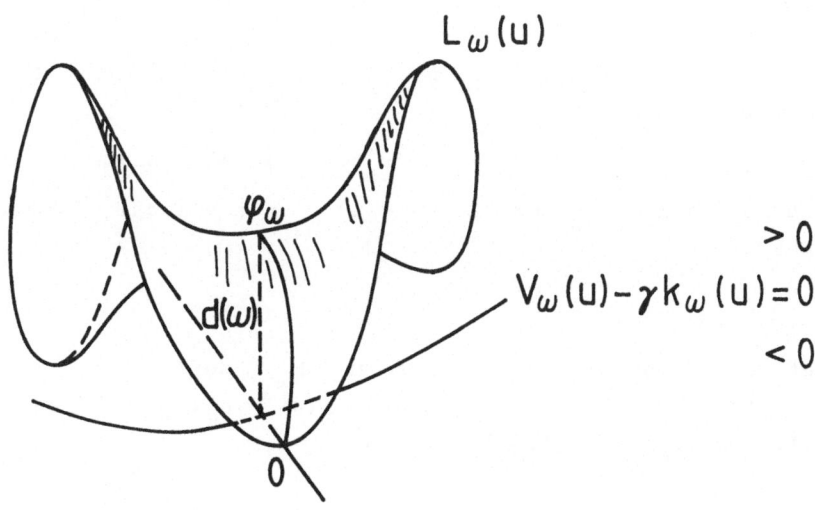

Having in mind this picture one has the most important informations necessary for the proofs of our main results:

Fix $\omega_o \in (\omega_1, \omega_2)$ and let $M_{\omega_o} \equiv \{u \in X | Q(u) = Q(\phi_{\omega_o})\}$.

<u>Theorem 1:</u> Let $d''(\omega_o) > 0$. Then the ϕ_{ω_o}-orbit is stable. In particular we have $Q(\phi_{\omega_o}) \neq 0$ and ϕ_{ω_o} is the local minimum of $E|_{\mu_{\omega_o}}$.

<u>Theorem 2:</u> Let $d''(\omega_o) < 0$. Then $E|_M$ is not locally minimized at ϕ_{ω_o} and the ϕ_{ω_o}-orbit is unstable.

Furthermore we have the following particular cases:

<u>Theorem 3:</u> If $Q(\phi_{\omega_o}) = 0$ then the ϕ_{ω_o}-orbit is unstable and ϕ_{ω_o} is not a local minimum of $E|_{M_{\omega_o}}$.

<u>Theorem 4:</u> If no symmetry U exists then any ground state ϕ_o of $E'(\phi) = 0$ is unstable. ϕ_o is not a local minimum of the energy.

Theorems 3/4 can be interpreted as abstract versions of Derrick's theorem [14] .

2. Outlines of the proofs

First of all we prove the following intermediate results.

<u>Theorem 2.1:</u>

a) Let ϕ_{ω_o} be a charged ground state $(Q(\phi_{\omega_o}) \neq 0)$. Then the energy E has a local minimum in M_{ω_o} at ϕ_{ω_o} if and only if $d(\omega)$ is convex at ω_o .

b) Let $Q(\phi_{\omega_o}) = 0$. Then ϕ_{ω_o} is not a local minimum of $E|_{\mu_\omega}$.

c) A ground state ϕ_o of $E'(\phi) = 0$ is not a local minimum of the energy.

<u>Proof:</u>

a) Let d be convex at ω_o . For $u \in M_\omega$ sufficiently close to ϕ_{ω_o} there exists ω such that $K_\omega(u) = K_\omega(\phi_\omega) = (1 - \gamma)d(\omega)$ since $d'(\omega_o) = -Q(\phi_{\omega_o}) \neq 0$. Thus

$$E(u) = L_\omega(u) + \omega\, Q(u)$$

$$\geq d(\omega) - \omega\, d'(\omega_o)$$

$$\geq d(\omega_o) - \omega_o\, d'(\omega_o) \qquad \text{by the convexity of } d$$

$$= E(\phi_{\omega_o}) \ .$$

For the opposite direction we define $\psi_\omega \equiv T_{\sigma(\omega)}\phi_\omega$ such that $\psi_\omega \in M_{\omega_o}$. Then

$$d(\omega_o) - \omega_o\, d'(\omega_o) = E(\phi_{\omega_o}) \leq E(\psi_\omega)$$

$$= \sigma^r\, K_\omega(\phi_\omega) - \sigma^s\, V_\omega(\phi_\omega) + \omega\, Q(\psi_\omega)$$

$$= (\sigma^r - \gamma\sigma^s)\, K_\omega(\phi_\omega) - \omega\, d'(\omega_o) \text{ by the}$$
$$\text{virial}$$
$$\text{theorem (1.13)}$$

$$\leq (1 - \gamma)\, K_\omega(\phi_\omega) - \omega\, d'(\omega_o)$$

$$= d(\omega) - \omega\, d'(\omega_o) \qquad \text{by (1.14)} \ .$$

For b) and c) we choose the curve $\psi_\sigma = T_\sigma\, \phi_{\omega_o}$ resp. $\psi_\sigma = T_\sigma\phi_\sigma$.

Now we sketch the stability proof, which is a _generalisation_ of the method presented by Shatah [11] for nonlinear Klein-Gordon equations:

Let $d''(\omega_o) > 0$. For arbitrary fixed ω we consider the modulated system

$$(H_{mod}) \qquad \frac{dv}{dt} = J\, L'_\omega(v) \ .$$

Then it can be shown that

$$(2.1.1) \qquad \mathbb{R}^1_\omega \equiv \{u \in X \,|\, L_\omega(u) < d(\omega)\, ,\, \gamma\, K_\omega(u) - V_\omega(u) > 0\} \cup \{0\}$$

$$(2.1.2) \qquad \mathbb{R}^2_\omega = \{u \in X \,|\, L_\omega(u) < d(\omega)\, ,\, \gamma\, K_\omega(u) - V_\omega(u) < 0\}$$

are invariant regions under the flow of (H_{mod}).

This fact can be understood if one has in mind the one-dimensional analogues presented in Section 1. If the energy L_ω is less then the height of the mountain pass one can never cross it. Here $\mathbb{R}^1_\omega$ represents the region inside the well and $\mathbb{R}^2_\omega$ represents the exterior region.

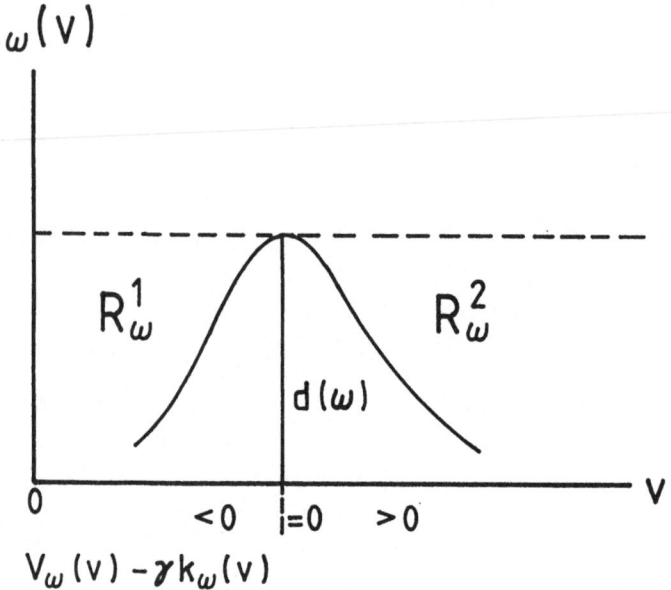

$$L_\omega(v)$$

$$R_\omega^1 \qquad R_\omega^2$$

$$d(\omega)$$

$$0 \qquad <0 \quad |=0 \quad >0 \qquad v$$

$$V_\omega(v) - \gamma k_\omega(v)$$

Furthermore we can characterize these regions as follows:

(2.2) $R_\omega^{1/2} = \{u \in X \mid L_\omega(u) < d(\omega) \mid (1-\gamma) \, K_\omega(u) \lessgtr d(\omega)\}$.

Now for every $\varepsilon > 0$ small enough there exists a $t > 0$ such that if $\|u - \phi_{\omega_o}\| < \delta$ then

(2.3) $d(\omega_+) < (1-\gamma) \, K_{\omega_o}(u) < d(\omega_-)$ if $Q(\phi_{\omega_o}) > 0$

or the reversed inequality if $Q(\phi_{\omega_o}) < 0$,   where $\omega_\pm = \omega \pm \varepsilon$

In addition $v_\pm$ defined by $u = U(\omega_\pm t) \, v_\pm$ satisfies

(2.4) $L_{\omega_\pm}(v_\pm) < d(\omega_\pm)$

which follows from the strict convexity of d in a neighborhood of ω_o .

Assume that ϕ_{ω_o} is unstable. Then there exists a sequence of initial data $u_n(o) \xrightarrow{\;\;} \phi_{\omega_o}$ in X and $\delta > 0$ and $t_n > 0$ such that

(2.5) $\|u_n(t_n) - \phi_{\omega_o}\| \geq \delta$.

Applying (2.3) we conclude

$$(1 - \gamma) \ K_{\omega_o}(u_n(t_n)) \xrightarrow{n \to \infty} d(\omega_o)$$

and (2.4) then implies

$$L_{\omega_o}(u_n(t_n)) \xrightarrow{n \to \infty} d^1 \leq d(\omega_o) \quad .$$

But now the assumptions on the minimization problem imply that there exists a subsequence of $(u_n(t_n^*))$ which tends to the ϕ_{ω_o}-orbit contradicting (2.5) .

The following figure illustrates the proof given above:

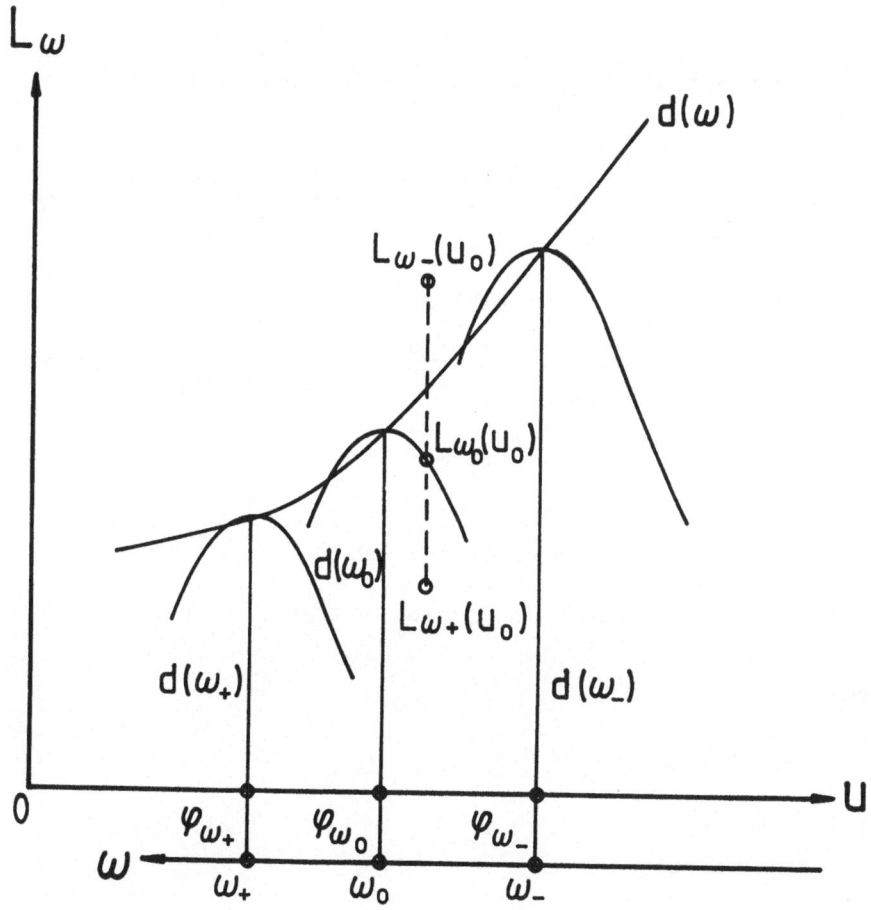

The instability proof first of all will be sketched for the case where we have no symmetry:

By Theorem 2.1 we know that ϕ_o (the ground state of $E'(u) = 0$) is a local maximum along the curve $\psi_\sigma = T_\sigma \phi_\sigma$.

We define $y = \dfrac{d}{d\sigma}\Big|_{\sigma=1} T_\sigma \phi_o$. There exists $Y \in X^*$ such that $JY = y$. We define a linear functional on X by

$$(2.6) \qquad A(u) \equiv - <Y,u> \quad .$$

Let $u_o \in U_\varepsilon(\phi_o)$. It suffices to show that $\dfrac{d\,A(u(t))}{dt}$ is bounded away from zero as long as $u(t) \in U_\varepsilon(\phi_o)$.

The trajectories along the vector field $- J A'(u)$ are given by

$$(2.7) \qquad \mathbb{R}(\lambda,v) = v + \lambda\, y \quad .$$

Then there exists a functional $\Lambda : U_\varepsilon \longrightarrow \mathbb{R}$ such that

$$(2.8) \qquad K(R(\Lambda(v),v)) = K(\phi_o) \quad .$$

Doing a Taylor expension for the energy and using the minimum property of ϕ_o we obtain for any $v \in U_\varepsilon(\phi_o)$ $(v \neq \phi_o)$

$$(2.9) \qquad E(\phi_o) < E(v) + \Lambda(v)\, P(v)$$

where $P(v) = <E'(v),y>$. Using $\dfrac{d\,A(u(t))}{dt} = P(u(t))$ as long as $u(t) \in U_\varepsilon$ we easily conclude.

In the presence of a symmetry we proceed in a similar way. Let $d''(\omega_o) < 0$ (or $Q(\phi_{\omega_o}) = 0$) .

The neighborhood U_ε is replaced by the tube

$$U_\varepsilon \equiv \{u \in X \mid \inf_{s \in \mathbb{R}} \|u - U(s)\phi_{\omega_o}\| < \varepsilon\} \quad .$$

We decompose

$$X = < U'(0)\, \phi_{\omega_o} > \oplus \tilde{X} \quad .$$

There exists a functional $s(u) : U_\varepsilon \longrightarrow \mathbb{R}$ such that $U(s(u))u - \phi_{\omega_o} \in \tilde{X}$. Then we define

$$(2.10) \qquad A(u) = - < Y_{\omega_o},U(s(u))u >$$

for any $u \in U_\varepsilon \cap M_{\omega_o}$ where $Y_{\omega_o} = J^{-1} \dfrac{d}{d\omega}\Big|_{\omega = \omega_o} \psi_\omega$ with ψ_ω the curve given in the proof of theorem 2.1. Up to simple modification the proof of instability will now run along the same lines as above .

3. Applications

3.1. We start with the logarithmic Schrödinger equation

(3.1) $i \phi_t + \Delta\phi + \phi \log |\phi|^2 = 0$ on $\mathbb{R}^n$, $n \geq 3$.

As mentioned in the introduction the ground state with frequency ω is of the form

$$\phi_\omega(x) = \phi_0(x) \exp(-\omega/2) \quad .$$

Hence

(3.2) $d(\omega) = \frac{1}{N} \exp(-\omega) \, K(\phi_0)$

with K given by (0.2a) and therefore all ground states are stable.

3.2. Consider the following nonlinear Schrödinger equation with a non-local nonlinearity

(3.3) $i \phi_t + \Delta\phi + \left(\int_{\mathbb{R}^N} V(x-y) \, |\phi(y)|^2 \, dy \right) \phi = 0$

As scaling group we take $T_\sigma \phi = \sigma\phi$. We define

(3.4a) $K_\omega(\phi) = \frac{1}{2} \int_{\mathbb{R}^N} |\nabla\phi|^2 + \omega|\phi|^2$, $\omega > 0$

(3.4b) $V_\omega(\phi) = \frac{1}{4} \int_{\mathbb{R}^N} \int_{\mathbb{R}^N} V(x-y) \, |\phi(y)|^2 \, |\phi(x)|^2 \, dxdy$

under suitable assumptions on V the assumptions of Section 1 are valid.

Consider, e.g. equation (3.3) on $\mathbb{R}^3$ with $V(x) = \frac{1}{|x|}$. This is the so-called Pekard-Choquard equation. The minimization problem was solved by Lieb [8].   Existence of solutions for the cauchy problem in $H_r^1(\mathbb{R}^3)$ was proved by Ginibre and Velo [6] .

Using the scaling properties of the stationary Pekard-Choquard we see that

$$\psi(x) = \omega^{-1} \phi_\omega(\omega^{-1/2} x)$$

is independent of ω . Hence

(3.5) $d(\omega) = \omega^{3/2} d(1)$

and therefore all ground states are stable. The same stability result
was obtained by Cazenave and Lions [15] using the concentration compactness principle.

3.3. Consider the logarithmic Klein-Gordon equation

(3.6) $\phi_{tt} - \Delta \phi = \phi \log (\phi)^2$ on $\mathbb{R}^N$, $n \geq 3$.

We study this problem in the space

$$X = W_r \oplus L^2_r (\mathbb{R}^N)$$

regarded as a real Banach space and W_r given by (0.3) resp. (0.4).

The energy and the charge are well-defined on X and given by

(3.7) $E(\phi_1,\phi_2) = \frac{1}{2} \int_{\mathbb{R}^n} |\phi_1|^2 + |\nabla\phi_2|^2 - \frac{1}{2} \int |\phi_2|^2 (-1+\log|\psi_2|^2)$

(3.8) $Q(\phi_1,\phi_2) = \text{Im} \int_{\mathbb{R}^n} \overline{\phi}_2 \phi_1$.

As scaling group we take $T_\sigma u(x) = u(\gamma/\sigma)$ where $u(x) = (\phi_1(x),\phi_2(x))^t$
and define

(3.9a) $K_\omega(\phi_1,\phi_2) = \frac{1}{2} \int_{\mathbb{R}^n} |\nabla\phi_2|^2$

(3.9b) $V_\omega(\phi_1,\phi_2) = \frac{1}{2} \int_{\mathbb{R}^n} |\phi_1|^2 - |\phi_2|^2 (-1+ \log |\phi_2|^2) + \omega \int_{\mathbb{R}^n} \phi_1\overline{\phi}_2$.

Now for any fixed $\omega \in \mathbb{R}$ we have the inequality

(3.10) $V_\omega(\phi_1,\phi_2) \leq V_\omega(i \omega \phi_2,\phi_2)$

with equality if and only if $\phi_1 = i \omega \phi_2$ and therefore the minimization problem yields indeed a solitary wave solution. For the logarithmic Klein Gordon equation the ground state is given by

(3.11) $\phi_\omega(x) = \phi_o(x) \exp(-\omega^2/2)$.

Hence $d(\omega) = \frac{1}{N} \exp(-\omega^2) \int_{\mathbb{R}^n} |\nabla\phi_o|^2$ and we have the following result:

If $\omega^2 \leq 1/2$ then the ground state ϕ_ω is unstable.

If $\omega^2 > 1/2$ it is stable.

3.4. We study the nonlinear wave equation

(3.12) $u_{tt} - \Delta u = g(u)$ on $\mathbb{R}^n$, $n \geq 3$

where for real-valued functions u on $\mathbb{R}^n$. Under suitable conditions
on $g(u)$ (see e.g. [1,9]) (3.12) possesses a nontrivial ground state
$U_0(x)$, U_0 is always unstable by Theorem 4.

4. Extensions and Comments

4.1. First of all it is natural to ask about the relation between the
model presented here and the one of Grillakis, Shatah and Strauss [7].

We have the following:

If E'' exists then $H_\omega = E''(\phi_\omega) - \omega\, Q''(\phi_\omega)$ where ϕ_ω is a ground
state satisfies the spectral assumptions of [7] , i.e. H_ω has
exactly one negative eigenvalue, a Kernel spanned by $U'(0)\phi_\omega$ and the
rest of the spectrum positive and bounded away from zero.

Indeed H_ω has at least some negative spectrum since $\frac{d^2}{d\sigma^2}\Big|_{\sigma=1}$,
$L_\omega(T_\sigma\phi_\omega) < 0$. On the other hand it has at most one
negative eigenvalue since ϕ_ω is a local minimum of L_ω on a C^1-hyper-
surface by (1.15) . The assumption for solution of the minimization
principle (1.10) .

4.2. An extension to non-scale covariant models is also possible. Let
L_ω satisfy the condition

(4.1) $< L'_\omega(\lambda u) - \lambda\, L'_\omega(u) , (\lambda - 1)\, u > \; < 0$ $u \neq 0$
                                                                                        $\lambda \neq 1$

and define

(4.2) $R_\omega(u) = < L'_\omega(u) , u >$.

We can define $d(\omega)$ by the minimization principle

(4.3) $d(\omega) = \inf \{L_\omega(u) , u \in X , u \neq 0 , R_\omega(u) = 0\}$.

Note that 0 is isolated in the set $\{u \in X , R_\omega(0) = 0$ in view of
condition (4.1) .

Under these assumptions one obtains the same criteria for stability/
instability in terms of $d(\omega)$ as for scale-covariant models.

4.3. The extension to more symmetries in the context of scale covariant functionals is very difficult. But there is one simple extension:

Let $K(u)$, $V(u)$ be invariant under N one parameter groups $(U_j(s))_{1 \leq j \leq N}$ which satisfy the 'commutation law'

$$(4.4) \qquad U_j(s) \ U_k(t) = U_k(t) \ U_j(s) \qquad \forall j_k \quad \forall s_t \in \mathbb{R} \ .$$

The associated conserved quantities Q_j are required to satisfy the following scaling properties:

Q_1 satisfies (1.7) while $Q_2, \ldots, Q_N$ may have different scaling behaviour. We look for ground states

$$(4.5) \qquad U_1(\omega_1 t) \ U_2(\omega_2 t) \ \ldots \ U_N(\omega_N t) \ \phi_{\vec{\omega}}$$

of (H) which satisfy $Q_2 = \ldots = Q_N = 0$.

In this case the stability of the orbit

$$\{U_1(s_1) \ \ldots \ U_N(s_N) \ \phi_{\vec{\omega}} \ , \quad s_j \in \mathbb{R} \}$$

is determined by the behaviour of the function

$$(4.6) \qquad d(\vec{\omega}) = E(\phi_{\vec{\omega}}) - \omega_1 \ Q(\phi_{\vec{\omega}})$$

in the variable ω_1 .

This assertion enables us to extend the stability results of Section 3 to stability under perturbations which are not necessarily radial. Consider e.g. the logarithmic Schrödinger equation (3.1) . Now we choose $X = W$ given by (0.3) instead of W_r . Then equation (3.1) is also invariant under translations. Gauge invariance and the of translation invariance satisfy the commutation law (4.4) .

The momentum is given by

$$(4.7) \qquad (u) = - \frac{1}{2} \int_{\mathbb{R}^n} \bar{u} \ \vec{\nabla} u$$

which satisfies the scale relation $\vec{P}(T_\sigma u) = \sigma^{n-1} \ \vec{P}(u)$.

Now ϕ_ω is not only a ground state in W_r , it is also a ground state in W . Furthermore it has zero momentum.

Hence by the result above its stability under perturbations in W is determined only by the frequency and there ϕ_ω is also stable in W .

References

[1] H. Berestycki and P.L. Lions, Arch. Rat. Mech. Anal. 82, 312 (1985)

[2] Ph. Blanchard, J. Stubbe and L. Vazquez, Ann. I.H.P. Phys. Théor. Vol. 47, 309 (1987)

[3] Ph. Blanchard, J. Stubbe and L. Vazquez, J. Phys. A 21, 1137 (1988)

[4] E. Brüning, I.H.E.S. preprint P86/53 (1986)

[5] Th. Cazenave, Nonlinear Analysis T.M.A. 7, 1127 (1983)

[6] J. Ginibre and G. Velo, Math. Zeitschrift 170, 109 (1980)

[7] M. Grillakis, J. Shatah and W.A. Strauss, J. Funct. Anal. 74, 160 (1987)

[8] E.H. Lieb, Stud. Appl. Math. 57, 93 (1977)

[9] P.L. Lions, Ann. I.H.P. Anal. Non Lin. 1, 109 (1984)

[10] L.E. Payne and D.H. Sattinger, Israel J. Math. 22, 273 (1975)

[11] J. Shatah, Comm. Math. Phys. 91, 313 (1983)

[12] J. Shatah and W.A. Strauss, Comm. Math. Phys. 100, 173 (1985)

[13] J. Stubbe, Diplomarbeit BI TP 86/10 (1986)

[14] J. Stubbe, Doktorarbeit, Universität Bielefeld, 1988

[15] Th. Cazenave, P.L. Lions, Comm. Math. Physics 85, 549 (1982)

A NOTE ON SOLUTIONS OF TWO-DIMENSIONAL SEMILINEAR ELLIPTIC VECTOR-FIELD EQUATIONS WITH STRONG NONLINEARITY

E. Brüning

Naturwissenschaftlich-theoretisches Zentrum
der Karl-Marx-Universität
Karl-Marx-Platz 10, DDR-7010 Leipzig
German Democratic Republic

and

Fakultät für Physik, Universität Bielefeld,
Universitätsstraße, 4800 Bielefeld 1, FRG

Abstract: We prove the existence of solutions u for some semilinear elliptic vector-field equations on $\mathrm{I\!R}^2$ with a nonlinearity which is allowed to grow at infinity 'nearly like a linear exponential' in $|u|$. In some cases a growth like $\exp(a|u|^{\gamma})$, $a > 0$, $0 < \gamma < 2$, is allowed.

This is achieved by introducing an appropriate space E of 'à priori-solutions' for which some important continuous imbeddings are proven replacing the well known Sobolev-imbeddings for the $d \geq 3$-dimensional case. Then the standard variational method is applied.

I. Introduction

A first important step in the existence proof of a solution of a semilinear elliptic equation of the type

$$- \Delta u = g(u)$$

$$u : \mathrm{I\!R}^d \to \mathrm{I\!R}^n , \quad g : \mathrm{I\!R}^n \to \mathrm{I\!R}^n , \quad n \geq 1 \qquad (1.1)$$

is always to fix a space E of à priori solutions. Motivated by the context where such equations occur one usually decides to look for solutions of finite kinetic energy

$$K(u) = \frac{1}{2} \|\nabla u\|_2^2 = \frac{1}{2} \int \sum_{j=1}^{n} \sum_{i=1}^{d} |\frac{\partial u_j}{\partial x_i}|^2 \, dx . \qquad (1.2)$$

Another natural contraint is that the functions in E should vanish at infinity in some sense. A way to express this is to define E to be the completion $E^{1,2}(\mathrm{I\!R}^d ; \mathbf{R}^n)$ of $C_0^\infty(\mathrm{I\!R}^d ; \mathbf{R}^n)$ with respect to the norm $f \to \|\nabla f\|_2$. For $d \geq 3$ this definition has proved to be rather useful (see for instance [3 - 5] and references there) because then by Sobolev's inequality

$$\| f \|_{2*} \leq S \| \nabla f \|_2 , \qquad 2^* = \frac{2d}{d-2} \tag{1.3}$$

this completion can be supposed to be realized as a subspace of $L^{2*}(\mathrm{IR}^d ; \mathrm{IR}^n)$. However for $d = 2$ this inequality breaks down and as a consequence this completion even is not a space of distributions as has been shown by J.L. Lions long ago.

So when treating the 'positive mass case' for scalar fields (i.e. $n = 1$) Berestycki et al. [2] decided to look for solutions in the subspace of radial functions of the Sobolev space $H^1(\mathrm{IR}^2)$. For the case of vector fields (i.e. $n > 1$) and mass zero the problem is considerably more complicated as is well known. This case has been treated by Brezis and Lieb [3]. They admit arbitrary polynomial bounds for g at infinity. The restriction on the behaviour of g near zero is expressed by the condition that the potential G of g is negative near zero (i.e. $G(0) = 0$ and $G(y) < 0$ for $0 < |y| < \varepsilon$ for some $\varepsilon > 0$). Their choice of the space of à priori solutions is well adapted to this situation but at the expense of not being a Banach space of functions on IR^2.

Furthermore by the context where such equations occur and also from a systematic point of view one should admit also a stronger than polynomial growth of g at infinity as it was done in [2]. Here we want to show that this is indeed possible even if a more general kind of behaviour of g near $y = 0$ is allowed than that treated in [2].

Basically we will follow the general strategy of [3] respectively of [5] but with a different space E of à priori solutions. Correspondingly the main step will be to introduce an appropriate space E and to study its properties, in particular properties of continuous imbeddings and continuity properties of certain classes of Niemytski operators on it. This is done in section II and finally applied in section III to solve equation (1.1) under appropriate conditions admitting also a growth of g at infinity nearly like a linear exponential in $|u|$ and in the 'positive mass case' like an exponential in $|u|^\gamma$, $0 < \gamma < 2$.

II. The Sobolev-type spaces $E_q(\mathrm{IR}^d ; \mathrm{IR}^n)$, $2 \leq d \leq q < \infty$, $n \geq 1$.

The spaces $E_q = E_q(\mathrm{IR}^d ; \mathrm{IR}^n)$ are defined as the completion of $C_0^\infty(\mathrm{IR}^d ; \mathrm{IR}^n)$ with respect to the norm

$$f \to \| f \| = \| f \|_q + \| \nabla f \|_d . \tag{2.1}$$

We consider this completion to be realized as a subspace of the Lebesgue space $L^q(\mathbb{R}^d ; \mathbb{R}^n)$ with norm $\| \cdot \|_q$.

If in section III for $d = 2$ such a space E_q is considered to be a space of à priori solutions for equation (1.1) the exponent q will be choosen according to the behaviour of the nonlinearity g near $y = 0$. Note that for $q = d = 2$ E_q is just the Sobolev space $H^1(\mathbb{R}^2 ; \mathbb{R}^n)$.

In this section we prove some properties of the spaces E_q and some important inequalities for elements in E_q which are fundamental for a variational approach to solutions of equation (1.1) (just as Sobolev's inequality is for the case $d \geq 3$).

Lemma 2.1

For any d, n, q, $2 \leq d \leq q < \infty$, $n \geq 1$, the spaces $E_q(\mathbb{R}^d ; \mathbb{R}^n)$ are real separable reflexive Banach spaces.

Proof: With minor changes the proofs of the corresponding statements for Sobolev spaces apply [6,7] .

The main information about the spaces E_q which we will use is contained in the following imbedding theorem.

Theorem 2.2

Suppose $d \leq q < \infty$ and denote $\frac{1}{p} = 1 - \frac{1}{d}$. Then we have

a) $E_q(\mathbb{R}^d) \hookrightarrow L^r(\mathbb{R}^d)$ for all $r \geq q$ with continuous injection, expressed by the inequality

$$\| u \|_r^r \leq C_r \| u \|_q^q \left(\frac{1}{d} \| \nabla u \|_d \right)^{r-q} \tag{2.2}$$

for all $u \in E_q$.

b) In particular the following inequality holds for all $u \in E_q$, all $k = 0, 1, 2, \ldots$,

$$\| u \|_{r(k)}^{r(k)} \leq (\alpha)_k^p \| u \|_q^q \left(\frac{1}{d} \| \nabla u \|_d \right)^{k \cdot p} \tag{2.3}$$

where $r(k) = q + k \cdot p$, $(\alpha)_k = \alpha(\alpha + 1) \ldots (\alpha + k - 1)$, $\alpha = 1 + \frac{q}{p}$.

Remark 2.1

Theorem 2.2 has some immediate consequences:

a) For $d = 2 \leq q' \leq q$ we have

$$H^1(\mathbb{R}^2 ; \mathbb{R}^n) = E_2 \subseteq E_{q'} \subseteq E_q$$

and by examples one sees that

$$E_{q'} \subsetneqq E_q$$

whenever $q' < q$.

b) For $q = d = 2$ Theorem 2.2 provides a simple proof of the well-known Sobolev imbeddings

$$H^1 \hookrightarrow L^r(\mathbb{R}^2) \qquad \text{for all } r \geq 2$$

with explicitly known imbedding constants. At the end of this section there is a comment on an impovement of these constants.

Proof: 1. Denote by $D^{1,1}$ the completion of $D = D(\mathbb{R}^d) = C_0^\infty(\mathbb{R}^d)$ with respect to the norm $\phi \longrightarrow \|\nabla \phi\|_1$ on D. By continuous extension the Gagliardo-Nirenberg inequality ([1,6] and references)

$$\|V\|_p \leq \frac{1}{d} \|\nabla V\|_1 , \quad \frac{1}{p} = 1 - \frac{1}{d} \tag{2.4}$$

still holds on all of $D^{1,1}$. By induction on $k = 0, 1, 2, \dots$ we will show that for any fixed $u \in E_q$

$$V_k = |u|^{\alpha+k} , \qquad \alpha = 1 + \frac{q}{p} \tag{2.5}$$

belongs to $D^{1,1}$ so that the above inequality applies. Then the theorem easily follows.

In order to prepare the induction proof we recall some facts about smooth approximations of Lebesgue-integrable functions:

Choose some functions $\chi, \eta \in D$ such that

$$0 \leq \eta \leq 1 \quad \text{supp } \chi, \eta \subseteq \{ x : |x| \leq 1 \} , \quad \int \chi \, dx = 1$$

and $\eta(x) = 1$ for all $|x| \leq \frac{1}{2}$.

Then define for $m = 1, 2, \dots$

$$\chi_m(x) = m^{-d}\chi(mx) \quad \text{and} \quad \eta_m(x) = \eta(\frac{1}{m} x) .$$

Clearly χ_m, $\eta_m \in D$, $\int \chi_m dx = 1$, and $\eta_m(x) = 1$ on $|x| \leq \frac{1}{2} m$.

Now for any $V \in L^s(\mathbb{R}^d)$ it is known [6,7] :

$$V * \chi_m \in C^\infty(\mathbb{R}^d) \quad \text{and} \quad V * \chi_m \longrightarrow V \quad \text{in} \quad L^s \qquad \text{a)}$$

$$\Psi_m = \eta_m(V * \chi_m) \in D(\mathbb{R}^d) \quad \text{and} \quad \Psi_m \longrightarrow V \quad \text{in} \quad L^s \qquad \text{b).}$$

(2.6)

2. Now suppose $u \in E_q$ to be given. Clearly $V_0 = |u|^\alpha$ belongs to $L^{\frac{q}{\alpha}}(\mathbb{R}^d)$ and according to the rules for weak derivatives we know

$$\nabla V_0 = \alpha |u|^{\alpha-1} \nabla |u| \quad, \quad \| \nabla |u| \|_d \leq \| \nabla u \|_d \quad .$$

Hence Hölder's inequality implies by choice of α

$$\| \nabla V_0 \|_1 \leq \alpha \| |u|^{\alpha-1} \|_p \; \| \nabla |u| \|_d \leq \alpha \| u \|_q^{\alpha-1} \| \nabla u \|_d$$

(2.7)

thus $\nabla V_0 \in L^1$.

Next we prove that V_0 can be approximated in the $\| \nabla \cdot \|_1$-norm by the functions

$$\Psi_m^0 = \eta_m(V_0 * \chi_m) \quad . \tag{2.8}$$

Let us write $\nabla V_0 - \nabla \Psi_m^0 = (1 - \eta_m)\nabla V_0 + \eta_m \cdot \{\nabla V_0 + \nabla V_0 * \chi_m\} - (\nabla \eta) \cdot (V_0 * \chi_m)$.
Then by Hölder's inequality and the fact $\| \eta_m \|_\infty \leq 1$ the following estimate is available

$$\| \nabla V_0 - \nabla \Psi_m^0 \|_1 \leq \| (1 - \eta_m) \nabla V_0 \|_1 + \| \nabla V_0 - \nabla V_0 * \chi_m \|_1 + \| \nabla \eta_m \|_\beta \| V_0 * \chi_m \|_{\frac{q}{\alpha}}$$

where $1 = \frac{1}{\beta} + \frac{\alpha}{q}$.

Since $\nabla V_0 \in L^1$, the first term tends to zero for $m \to \infty$ by dominated convergence. The limit $m \to \infty$ of the second term also vanishes by statement 2.6.a).

Since $\| V_0 * \chi_m \|_{\frac{q}{\alpha}}$ is bounded and since

$$\| \nabla \eta_m \|_\beta = m^{\frac{d}{\beta} - 1} \| \nabla \eta \|_\beta \quad , \quad \frac{d}{\beta} - 1 = -\frac{d}{q}$$

the limit $m \to \infty$ of the third term also vanishes.

This proves $\nabla\psi_m^0 \to \nabla V_0$ in L^1 and thus $V_0 \in D^{1,1}$.
Hence inequality (2.4) applies

$$\| V_0 \|_p \le \frac{1}{d} \| \nabla V_0 \|_1$$

that is by inequality (2.7)

$$\| u \|_{\alpha \cdot p}^\alpha \le \frac{\alpha}{d} \| u \|_q^{\frac{q}{p}} \| \nabla u \|_d .$$

Now we have $\alpha p = q + p = r(1)$. Therefore inequality (2.3) holds for $k = 1$.

3. Our induction hypothesis is that for some $k \ge 1$

$$V_j = |u|^{\alpha+j} \in D^{1,1} \qquad\qquad \text{a)}$$

$$\text{and}\qquad \| V_j \|_p = \| u \|_{p(\alpha+j)}^{\alpha+j} \le (\alpha)_{j+1}^p \| u \|_q^{\frac{q}{p}} \left(\frac{1}{d}\| \nabla u \|_d\right)^{j+1} \quad \text{b)}$$

$$\text{(2.9)}$$

for $j = 0, 1, \ldots k-1$.

For V_k it follows $V_k \in L^{q_k}(\mathbb{R}^d)$ with

$$q_k = \frac{p(\alpha+k-1)}{\alpha+k} > 1$$

and for ∇V_k we get

$$\nabla V_k = (\alpha+k) |u|^{\alpha+k-1} \nabla|u| = (\alpha+k) V_{k-1} \nabla|u|$$

hence by induction hypothesis $\nabla V_k \in L^1$ follows from Hölder's inequality.

Now we proceed as for $k = 0$. This time the smooth approximation are

$$\psi_m^k = \eta_m \cdot (V_k * \chi_m) , \qquad m = 1, 2, \ldots$$

and the relevant estimate is

$$\| \nabla V_k - \nabla\psi_m^k \|_1 \le \| (1 - \eta_m) \nabla V_k \|_1 + \| \nabla V_k - \chi_m * \nabla V_k \|_1 + \| \nabla \eta_m \|_{q_k'} \| V_k * \chi_m \|_{q_k}$$

and we conclude as above since

$$\| \nabla \eta_m \|_{q_k'} = m^{\frac{d}{q_k'} - 1} \| \nabla \eta \|_{q_k'}$$

with $\dfrac{d}{q_k'} - 1 = d\left\{ \dfrac{1}{q_k'} - \dfrac{1}{d} \right\} = d \cdot \left\{ \dfrac{1}{p} - \dfrac{1}{q_k} \right\} = -\dfrac{d}{r(k)} < 0$.

This proves $V_k \in D^{1,1}$ so that by the Gagliardo-Nirenberg inequality

$$\| V_k \|_p \leq \frac{1}{d} \| \nabla V_k \|_1 = \frac{\alpha + k}{d} \| V_{k-1} \nabla |u| \|_1 \leq \frac{\alpha + k}{d} \| V_{k-1} \|_p \| \nabla u \|_d$$

follows. Thus inequality (2.9.b) holds also for $j = k$.

Hence this inequality and the statement $V_k \in D^{1,1}$ hold for all $k = 0, 1, 2, \ldots$. Therefore the inequality in part b) of the theorem follows. The inequality in part a) follows from this using the interpolation inequality where the constant C_r is explicitly known.

The following two corollaries provide some information about the definition and some continuity properties of certain Niemytski operators on E_q .

Corollary 2.3

Suppose F is a continuous function $\mathbb{R}^n \to \mathbb{R}$ such that

$$|F(y)| \leq C |y|^q \sum_{k=0}^{\infty} a_k |y|^{\frac{p}{s} k} \quad \text{for all} \quad y \in \mathbb{R}^n \quad (2.10)$$

with some exponents $d \leq q < \infty$, $1 \leq s < \infty$, and some coefficients a_k satisfying

$$\limsup_{k \to \infty} | a_k(\alpha)_k^{\frac{p}{s}} |^{1/k} = \frac{1}{R} < \infty . \quad (2.11)$$

Then the Niemytski-operator $\hat{F}_s$ associated with F maps the subset $\left\{ u \in E_q (\mathbb{R}^d ; \mathbb{R}^n) \mid \frac{1}{d} \| \nabla u \|_d < R^{\frac{p}{p}} \right\} = Z_R$ into $L^1(\mathbb{R}^d)$ and satisfies

$$\| \hat{F}(u) \|_1 \leq C \| u \|_q^q \sum_{k=0}^{\infty} b_k [\frac{1}{d} \| \nabla u \|_d]^{\frac{p}{s} k} < \infty , \quad (2.12)$$

where $\quad b_k = a_k(\alpha)_k^{\frac{p}{s}}$. $\quad (2.13)$

Proof: For $u \in Z_R$ the following estimate is available by Theorem 2.2

$$\|\hat{F}(u)\|_1 = \int |F(u(x))| dx \leq C \int \sum_{k=0}^{\infty} a_k |u|^{q+\frac{p}{s}k} dx \leq$$

$$\leq C \sum_{k=0}^{\infty} a_k \| |u|^{\frac{q}{s}} \cdot |u|^{\frac{r(k)}{s}} \|_1 \leq C\|u\|_q^{\frac{q}{s}} \sum_{k=0}^{\infty} a_k \|u\|_{r(k)}^{\frac{r(k)}{s}}$$

$$\leq \|u\|_q^q C \sum_{k=0}^{\infty} b_k [\frac{1}{d} \| \nabla u\|_d]^{\frac{p}{s}k} .$$

Since by assumption $[\frac{1}{d} \| \nabla u\|_d]^{\frac{p}{s}} < R$ this series converges, this poves the collary.

Corollary 2.4

Under the assumptions of Corollary 2.4 with $R = +\infty$ the Niemytski operator $\hat{F}$ is sequential continuous as a map from E_q equipped with weak topology into L^1_{loc}, i.e. if $u_i \to u$ weakly in E_q then $\hat{F}(u_i) \to \hat{F}(u)$ in $L^1_{loc}(\mathbb{R}^d)$.

Proof: If a sequence $(u_i)_{i \in N}$ converges weakly in E_q to some element u then it is (strongly) bounded

$$\sup_i \| u_i \| = C < \infty$$

and we may assume also $u_i \to u$ almost everywhere on $\mathbb{R}^d$. By continuity of F it follows $\hat{F}(u_i) \to \hat{F}(u)$ for $i \to \infty$ for almost every $x \in \mathbb{R}^d$. Thus the statement follows from Vitali's convergence theorem if we can show

$$\sup_i \| \pi_A \hat{F}(u_i)\|_1 = H(|A|) \to 0 \quad \text{for} \quad |A| \to 0$$

for any measurable subset $A \subset \mathbb{R}^d$, $|A| < \infty$, π_A the operator of multiplication by 1_A. Similarly as in the proof of Corollary 2.3 we have for such a set A

$$\| \pi_A \hat{F}(u_i)\|_1 \leq \| \pi_A u_i \|_q^{\frac{q}{s}} C \sum_{k=0}^{\infty} a_k \|u_i\|_{r(k)}^{\frac{r(k)}{s}} .$$

If we observe now that with $\frac{1}{\beta} = \frac{1}{q} - \frac{1}{q+p} > 0$

$$\|\pi_A u_i \|_q \leq |A|^{1/\beta}\|u_i \|_{r(1)}$$

we get

$$\|\pi_A \hat{F}(u_i)\|_1 \leq |A|^{\frac{q}{s'\beta}} \|u_i\|_{r(1)}^{\frac{q}{sy}} \|u_i\|_q^{\frac{q}{s}} \sum_{k=0}^{\infty} b_k \left(\frac{1}{d} \|\nabla u_i\|_d \right)^{\frac{p}{s}k}$$

and thus by boundedness of $(u_i)_{i \in \mathbb{N}}$ in E_q and by Theorem 2.2

$$\sup_i \|\pi_A \hat{F}(u_i)\|_1 \leq |A|^{\frac{q}{s'\beta}} C \quad,$$

since the series converges everywhere. Thus we conclude.

Some examples will illustrate the growth restrictions used in the above corollaries.

Example 1

Suppose $F : \mathbb{R}^n \to \mathbb{R}$ to be continuous satisfying

$$|F(y)| \leq (a|y|^\gamma)^{n_\gamma} \sum_{k=0}^{\infty} \frac{a^k |y|^{\gamma k}}{(n_\gamma + k)!} = e^{a|y|^\gamma} - \sum_{l=0}^{n_\gamma - 1} \frac{(a|y|^\gamma)^l}{l!}$$

with $0 < \gamma \leq 1$, $a > 0$ and $q = \gamma n_\gamma \geq d$.

Example 2

The continuous function $F : \mathbb{R}^n \to \mathbb{R}$ is supposed to satisfy for some constants $a, b > 0$ and some exponents $0 < \gamma \leq 1$, $d \leq q < \infty$

$$|F(y)| \leq b|y|^q \operatorname{ch}(a|y|^\gamma) \ .$$

In both cases we get for the radius R_γ of convergence for the series defined by the coefficients b_k according to eq.(2.13):

$$R_\gamma = \begin{cases} \infty & \text{for} \quad 0 < \gamma < 1 \\ \text{finite} & \text{for} \quad \gamma = 1 \\ 0 & \text{for} \quad \gamma > 1 \ . \end{cases}$$

Hence these corollaries are conviently applied for $0 < \gamma < 1$, i.e. if F is bounded at infinite by some exponential of the form

$$e^{a|y|^\gamma} \ .$$

For the case of a linear exponential (i.e. $\gamma = 1$) some more complications arise. This case is not yet worked out in detail.

If we compare the estimates given in Theorem 2.2 with the corresponding estimates for the case $q = 2$ obtained by other methods we see that at least for $q = 2$ our estimates are not the best possible one's. We comment on this point.

For $q = 2$ the space E_q coincides with the usual Sobolev space $H^1 = H^1(\mathbb{R}^2 ; \mathbb{R}^n)$. It is well known that the Fouriertransform can be used to characterize conveniently elements of L^p-spaces by integrability conditions on their Fouriertransforms only for $p = 2$. Relying on Theorem 5.3 of reference [9] this fact can be used to give a simple proof of the following inequalities

$$(2.14) \qquad \|u\|_r \leq \alpha(r) \|u\|_{1,2} \quad u \in H^1 , \quad \|u\|_{1,2}^2 = \|u\|_2^2 + \|\nabla u\|_2^2$$

for all $r \geq 2$ with

$$\alpha(r) = 2^{-3(\frac{1}{2} - \frac{1}{r})} \pi^{-\frac{1}{2} + \frac{3}{r}} (r - 2)^{(\frac{1}{2} - \frac{1}{r})} , \qquad (2.15)$$

hence $\alpha(r) \leq c\, r^{1/2}$ with some constant $c > 0$ while inequality (2.3) says

$$\|u\|_r \leq \text{const } r \|u\|_{1,2} \qquad (2.16)$$

which is a weaker estimate.

Therefore one expects that the estimates of Theorem 2.2 can be improved also for the cases $q > 2$. However for $q > 2$ the convenient Fouriertransform method is not available. So a more general approach has to be used, may be along the lines of the method used for the Moser-Trudinger inequility [1,6,8].

We consider some examples. For $q = 2$, $a > 0$, and $0 < \gamma \leq 2$ choose

$$F(y) = e^{a|y|^\gamma} - \sum_{k=0}^{n_\gamma - 1} \frac{(a|y|^\gamma)^k}{k!} \qquad (2.17)$$

with $\gamma \cdot n_\gamma \geq 2$. Relying on (2.14) - (2.15) we get for all $u \in H^1$

$$\| \hat{F}(u) \|_1 \leq \sum_{k=n_\gamma}^{\infty} \frac{a^k}{k!} \alpha(\gamma k)^{\gamma k} \| u \|_{1,2}^{\gamma k}$$

and we see, using definition 2.11, that the radius R of convergence satisfies

$$R = R_\gamma = \begin{cases} \infty & \text{for} & 0 < \gamma < 2 \\ \text{finite} & \text{for} & \gamma = 2 \\ 0 & \text{for} & \gamma > 2 \end{cases} \qquad (2.18)$$

We mention another example:

$$F(y) = |y|^q \, \text{ch} \, a|y|^\gamma \ , \quad a > 0 \ , \quad q \geq 2 \ , \quad 0 < \gamma \leq 2 \ . \quad (2.19)$$

For $q = 2$ we can proceed as above to obtain (2.18) also in this case. For $q > 2$ however when we rely on Corollary 2.3 we get according to our example 2 from above for $\gamma = 1$ and all $\beta > 0$ with $A = \beta \frac{a}{2} (1 + \frac{q}{2}) < 1$, $u \in E_q$, $u \neq 0$:

$$(2.20) \int_{\mathrm{IR}^2} |u|^q \left\{ e^{a\beta \left(\frac{|u|}{\| \nabla u \|_2} \right)^\gamma} + e^{-a\beta \left(\frac{|u|}{\| \nabla u \|_2} \right)^\gamma} \right\} dx \leq 2 \| u \|_q^q (1 - A)^{-1}$$

while for $0 < \gamma < 1$ the integral is finite for all $\beta > 0$ since then $R_\gamma = \infty$.

III. Solution of some two dimensional vector field equations

As we will see the class of spaces E_q introduced in the last section is well adapted to look for solution of equation (1.1) in them for $d = 2$. According to the behaviour of the nonlinearity near $y = 0$ the exponent $q \geq 2$ has to be fixed.

So first we list our hypotheses on the nonlinearity g :

(H_0) $g : \mathrm{IR}^n \to \mathrm{IR}^n$ is continuous , $g(0) = 0$, such that

         $g(y) = \text{grad } G(y)$ for $y \neq 0$ for some potential G .

(H_1) $G : \mathbb{R}^n \to \mathbb{R}$ is continuous, of class C^1 in $\mathbb{R}^n \smallsetminus \{0\}$,

$G(0) = 0$, $G(y) > 0$ somewhere.

(H_2) G admits a decomposition $G = G_+ - G_-$ with continuous non-

negative functions $G_\pm$, $G_\pm(0) = 0$, such that there exists

an exponent $q \geq 2$ and a constant $b > 0$ with

(i) $b|y|^q \leq G_-(y)$ for all $y \in \mathbb{R}^n$

(ii) $G_+(y) \leq o(|y|^q)$ for $|y| \to 0$.

(H_3) $y \mapsto |y||g(y)|$ satisfies the growth restrictions as expressed

by statements (2.10) and (2.11) with $R = + \infty$.

(H_3') $G(\cdot)$ satisfies the growth restrictions of assumption (H_3) .

Remarks

a) According to (H_2) the potential G is negative near $y = 0$,
   more precisely $G(y) \leq -c|y|^q$ for all sufficiently small $|y|$
   with some constant $0 < c < b$. Hence this assumption is a particular case of the corresponding assumption in [3] .

b) By the examples mentioned in section II we know that assumption
   (H_3') admits potentials G which grow nearly like a linear exponential. More precisely (H_3') allows

$$|G(y)| \leq c |y|^q e^{a|y|^\gamma}   \text{ for } |y| \to \infty$$

with some constants $a, c > 0$, and some exponent $0 < \gamma < 1$.
For $q = 2$ we can allow $0 < \gamma < 2$ according to our remarks in
the previous section. Hence this assumption is considerably more
general then the corresponding one of a polynomial bound.

According to these assumptions we decide to look for solutions in the
Banach space $E_q = E_q(\mathbb{R}^2 ; \mathbb{R}^n)$ where the exponent $q \geq 2$ is given
by assumption (H_2) .

When looking for weak solutions of equation (1.1) one clearly has to
ensure that the 'potential' is differentiable in some sense. It turns
out that essentially the weakest notion of differentiability which is
natural in this context is sufficient for our purposes. So a first
lemma states the existence of linear continuous Gâteaux-derivatives in
all directions $v \in C_0^\infty (\mathbb{R}^2 ; \mathbb{R}^n)$ for all points where it is defined.

Lemma 3.1

Under the assumptions (H_0), (H_1), and (H_3) the functional

$$V(u) : = \int G(u(x)) \, dx , \qquad u \in D(V)$$

$$D(V) = \{v \in E_q | \ G(u(\cdot)) \in L^1(\mathbb{R}^2)\} \tag{3.1}$$

has linear continuous Gâteaux-derivatives $V'(u;v)$ at every $u \in D(V)$
in all directions $v \in C_0^\infty (\mathbb{R}^2 ; \mathbb{R}^n)$ given by

$$V'(u;v) = \int g(u) \cdot v \, d x . \tag{3.2}$$

Proof: The standard versions of such a differentiability result assume
some polynomial bound for g and G (see [4,5] and references there).
So they don't apply directly. However if we take into account the basic
inequalities from Theorem 2.2 it is not hard to show that the proof
of Proposition 2.5 in [4] can be extended to the present case.
The estimates for the term

$$\| \hat{g} (u) \cdot V [\lambda \leq |u|] \|_1$$

$([\lambda \leq |u|] = $ characteristic function of the set $\{x \in \mathbb{R}^2 | \lambda \leq |u(x)|\})$
used there can be done in analogy with the estimate for

$$\| \hat{G}_+(f_j)[\lambda < |f_j|] \|_1$$

for $\lambda \to \infty$ which will be given explicitly later in the proof of Theorem 3.3.

Lemma 3.2

Suppose g satisfies (H_0), (H_1), (H_3), and

$$g^{-1}(0) \subseteq \{0\} \cup \{y \in \mathbb{R}^n \mid \delta \leq |y|\} \tag{3.3}$$

for some $\delta > 0$. If for $u \in E_q$

$$\hat{g}(u) = 0$$

holds in the sense of distributions then the function u vanishes :
u = 0 .

Proof: This results follows from the observation that weakly differentiable functions have no finite jumps [3] . Relying on [3'] a complete proof is given in [5] . By our assumptions on g we know
$\hat{g}(u) \in L^1_{loc}$ for every $u \in E_q$, hence $\hat{g}(u) = 0$ almost everywhere on
$\mathbb{R}^2$. By the assumption about the zeros of g this equation can hold
only if $|u|$ had a jump of height at least δ . Thus u = 0 follows.

Remark 3.1

a) Assumptions (H_0), (H_1), and (H_3) imply assumption (H_3') .

b) Hypotheses (H_0) - (H_2) imply assumption (3.3) .

Theorem 3.3

If the assumption (H_0) - (H_2) and (H_3) are satisfied then the equation (1.1) has a nontrivial solution in E_q.

Proof: Step 1: Construction of an appropriate minimizing sequence:
By our assumptions and definition (3.1) the following minimization
problem is well defined:

$$I = \inf \{K(v) = \frac{1}{2} \|\nabla v\|_2^2 \mid v \in D(V), v \neq 0, V(v) \geq 0\}$$

and there is a minimizing sequence

$$f_j \in D(V), \; f_j \neq 0, \; V(f_j) \geq 0, \; I = \lim_{j \to \infty} K(f_j). \tag{3.5}$$

Because of the covariance properties

$$K(v_\sigma) = K(v), \; V(v_\sigma) = \sigma^{-2} V(v), \; v_\sigma(\cdot) = v(\sigma \cdot) \tag{3.6}$$

we may assume in addition for all $\; j \in \mathbb{N}$

$$\| f_j \|_q = 1 . \tag{3.7}$$

Hence we get the following chain of inequalities using first part i) and then part ii) of assumption (H_2) :

$$b \leq \int \hat{G}_-(f_j)dx \leq \int \hat{G}_+(f_j)dx \leq$$

$$\leq \int [|f_j|<\varepsilon]\hat{G}_+(f_j)dx + \int [\varepsilon \leq |f_j| \leq \lambda]\hat{G}_+(f_j)dx + \int[\lambda<|f_j|]\hat{G}_+(f_j)dx$$

$$\leq a(\varepsilon) + C_{\varepsilon,\lambda} \; |[\varepsilon \leq |f_j| \leq \lambda]| + J_\lambda(f_j) \tag{3.8}$$

where by part (ii)

$$G_+(y) \leq a(\varepsilon) \; |y|^q \qquad \text{for} \qquad |y| < \varepsilon$$

with $\qquad a(\varepsilon) \to 0 \qquad$ for $\quad \varepsilon \to 0 \quad$ and where

$$C_{\varepsilon,\lambda} = \sup \{G_+(y)| \cdot \varepsilon \leq |y| \leq \lambda\} < \infty .$$

The Lebesgue measure of a measurable set $\;M\;$ is denoted by $\;|M|$. The third term in (3.8) is controlled by assumption (H_3):

$$\int [\lambda<|f_j|] \; \hat{G}_+(f_j)dx \leq \int([\lambda<|f_j|] \sum_{k=0}^{\infty} a_k(f_j)^{q+\frac{2}{s}k}) \; dx$$

$$\leq \sum_{k=0}^{\infty} a_k \; \|[\lambda<|f_j|]|f_j|^{\frac{q}{s}}|f_j|^{\frac{r(k)}{s}} \|_1 \leq$$

$$\leq \; \|[\lambda<|f_j|]f_j \|_q^{\frac{q}{s}} \cdot \sum_{k=0}^{\infty} a_k \|f_j \|_{r(k)}^{\frac{r(k)}{s}} .$$

By Theorem 2.2 we have

$$\| f_j \|_{r(k)}^{r(k)} \leq (\alpha)_k^2 \left(\frac{1}{2} K(f_j) \right)^k \tag{3.9}$$

where $\alpha = 1 + q/2$, hence

$$J_\lambda(f_j) \leq \| [\lambda < |f_j|] f_j \|_q^{\frac{q}{s}} \sum_{k=0}^\infty b_k \left(\frac{1}{2} K(f_j) \right)^{\frac{2}{k}} \tag{3.10}$$

with b_k according to definition (2.13) .

Since $(K(f_j))_{j \in \mathbb{N}}$ is bounded Theorem 2.2 implies that $(\| f_j \|_r)_{j \in \mathbb{N}}$ is bounded for every $r \geq q$.

Therefore for fixed $r > q, \frac{1}{q} = \frac{1}{r} + \frac{1}{\tau}$, we know

$$\| [\lambda < |f_j|] f_j \|_q \leq |[\lambda < |f_j|]|^{1/\tau} \| f_j \|_r \leq C |[\lambda < |f_j|]|^{1/\tau} \leq C' \lambda^{-\frac{q}{\tau}}$$

since $1 = \| f_j \|_q^q \geq \| [\lambda < |f_j|] |f_j|^q \|_1 \geq \lambda^q |[\lambda < |f_j|]|$.

By assumption (H_3) the series in inequality (3.10) converges everywhere on $\mathbb{C}$. Thus we get

$$\sup_j J_\lambda(f_j) \longrightarrow 0 \quad \text{for} \quad \lambda \longrightarrow \infty . \tag{3.11}$$

Hence there are $0 < \epsilon < \lambda < \infty$ such that by inequality (3.8)

$$\frac{b}{2} \leq C_{\epsilon,\lambda} |[\epsilon \leq |f_j| \leq \lambda]|$$

holds for every $j \in \mathbb{N}$. Now we can use the lemma of concentration by translation [3,5] to conclude as in [3] that there is a minimizing sequence $(u_j)_{j \in \mathbb{N}}$ and an element $u \in E_q$ such that

$$u_j \to u \quad \text{for} \quad j \to \infty .$$

   a) weakly in E_q

   b) almost everywhere on $\mathbb{R}^2$ (3.12)

   c) $u \neq 0$.

Step 2: We are going to show that the limit function u according to
statement (3.12) actually is a solution of the differential equation
(1.1).

Inequality (3.8) implies in particular

$$\sup_{i \in \mathbb{N}} \int \hat{G}_+(u_i) \, dx \; < \; \infty \; .$$

Thus by (3.12.b) and Fatou's lemma we get $\hat{G}_+(u) \in L^1(\mathbb{R}^2)$.
By $V(u_j) = V_+(u_j) - V_-(u_j) \geq 0$ we get in the same way $\hat{G}_-(u) \in L^1(\mathbb{R}^2)$.
This proves $\hat{G}(u) \in L^1$, i.e. $u \in D(V)$.
If also $V(u) = \int \hat{G}(u) dx \geq 0$ were known the limit function u would
be a non trivial minimizer. In the case of scalar fields (n = 1) it
is easy to prove $V(u) \geq 0$ (see later remark). Here instead it is
shown directly that a suitably scaled version of u is a weak solution
of our equation (without showing first $V(u) \geq 0$).

Take a fixed $v \in C_0^\infty (\mathbb{R}^2 ; \mathbb{R}^n)$ with compact support K and apply
Corollary 2.4 to the sequences $(\pi_K u_i)_{i \in \mathbb{N}}$ and $(\pi_K u_i + v)_{i \in \mathbb{N}}$.

Since $\pi_K \hat{G}(u_i) = \hat{G}(\pi_K u_i)$ this corollary implies
that $\hat{G}(\pi_K u_i) \to \hat{G}(\pi_K u)$ in $L^1(\mathbb{R}^2)$ and similarly

$$\hat{G}(\pi_K u_i + v) \to \hat{G}(\pi_K u + v), \quad \text{for } i \to \infty ;$$

hence

$$\hat{G}(\pi_K u_i + v) - \hat{G}(\pi_K u_i) \xrightarrow[i \to \infty]{} \hat{G}(\pi_K u + v) - \hat{G}(\pi_K u) \quad \text{in } L^1 \quad (3.13)$$

but $\hat{G}(\pi_K u + v) - \hat{G}(\pi_K u) = \hat{G}(u + v) - \hat{G}(v) ,$

and therefore (3.13) proves

$$V(u_i + v) - V(u_i) \xrightarrow[i \to \infty]{} V(u + v) - V(u) . \qquad (3.14)$$

Now using the differentiability result of Lemma 3.1 one can show just
as in [3]   that there is some $\lambda \geq 0$ such that

$$V'(u;v) = \lambda K'(u;v) \quad \text{for for all} \quad v \in C_0^\infty (\mathbb{R}^2 ; \mathbb{R}^n) . \qquad (3.15)$$

If $\lambda = 0$ then Remark 3.1b and Lemma 3.2 imply $u = 0$.

This contradiction proves $\lambda > 0$ and we may rescale u according to

$$\overline{u} = u_\sigma, \quad \sigma = \sqrt{\lambda}$$

and obtain a nontrivial weak solution $\overline{u}$:

$$V'(\overline{u};v) = K'(\overline{u};v) \qquad \text{for all} \quad v \in C_0^\infty. \qquad (3.16)$$

Remark 3.2

a) In the case of scalar fields $(n = 1)$ and an even potential G spherically symmetric rearrangement of functions can be used. Then it suffices to restrict to the subspace of spherically symmetric (nonincreasing) functions in E_q.

In this case for $q = 2$ (the 'positive mass' case) the existence of infinitely many solutions has been stated in [2] for a class of potentials which are bounded at infinity by

$$e^{a|y|^\gamma}, \quad a > 0, \quad \gamma = 2. \qquad (3.17)$$

Taking into account the comments on the case $q = 2$ at the end of section 2 it is not hard to see by (2.14) - (2.15) that Theorem 3.3 can easily be extended in the vector field case for $q = 2$ to cover a nonlinearity satisfying (3.17) with $0 < \gamma < 2$. For $\gamma = 2$ the radius of convergence of the relevant power series is finite according to (2.18); this allows only to treat those cases for which it is possible to find a minimizing sequence of elements with sufficiently small H^1-norms. This point is still under investigation.

b) Recall from [6], Theorem 8.8 that the basic step of the elliptic regularity theory applies if $\hat{g}(u) \in L^2_{loc}$ is known. The following Lemma provides this information.

Lemma

If the continuous function g satisfies hypothesis (H_3) then for any $u \in E_q(\mathbb{R}^2) : \hat{g}(u) \in L^2(\mathbb{R}^2)$.

Proof: By (H_3) one has

$$\int_{\mathbb{R}^2} |\hat{g}(u)|^2 \, dx \leq \sum_{k,l=0}^{\infty} a_k \, a_l \, \| \, |u|^{2(q-1) + \frac{2k}{s} + \frac{2l}{s}} \|_1 \, .$$

If $q \geq 2$ then $q' = 2(q-1) \geq q$ so that we can apply the estimates of Theorem 2.2 for $r'_k = q' + 2k$, i.e. $r'_{2k} = 2(r_k - 1)$, where $r_k = r(k) = q + 2k$. Hence with $\alpha' = 1 + \frac{q'}{2} = q$ we have

$$\| u \|_{r'_{2k}}^{r'_{2k}} \leq (\alpha')_{2k}^2 \, \|u\|_{q'}^{q'} \left(\tfrac{1}{2} \, \|\nabla u\|_2 \right)^{2(2k)}$$

and thus by Hölder's inequality and Theorem 2.2 (observe $E_q \subseteq E_{q'}$ for $q \leq q'$)

$$\| \, |u|^{2(q-1) + \frac{2k}{s} + \frac{2l}{s}} \|_1 \leq \| \, |u|^{q-1+\frac{2}{s} k} \|_2 \| \, |u|^{q-1+\frac{2l}{s}} \|_2$$

$$\leq \|u\|_{q'}^{q'/s} \, \|u\|_{r'_{2k}}^{r'_{2k}/s} \, \|u\|_{r'_{2l}}^{r'_{2l}/s} \leq \|u\|_{q'}^{q'} (\alpha')_{2k}^{1/s} (\alpha')_{2l}^{1/s} A^{k+l}$$

where $A = \left(\tfrac{1}{2} \, \|\nabla u\|_2 \right)^{2/s}$.

This proves finally

$$\| \hat{g}(u)\|_2 \leq \|u\|_{2(q-1)}^{q-1} \cdot \sum_{k=0}^{\infty} a_k (q)_{2k}^{1/s} A^k < \infty \, ,$$

since

$$\limsup_{k \to \infty} |a_k(\alpha)_k^{2/s}|^{1/k} = 0$$

implies

$$\limsup_{k \to \infty} |a_k(q)_{2k}^{1/s}|^{1/k} = 0$$

so that the above series converges for every $A \in \mathbb{C}$.

Therefore by Theorem 8.8 of [6] it follows that a weak solution u of equation (1.1) in E_q actually belongs to $W_{loc}^{2,2} \cap E_q$.

Hence such a solution solves this equation in the sense of equality almost everywhere on $\mathbb{R}^2$.

c) In analogy with method explained in [4] for the $d \geq 3$-dimensional case equations of the form

$$- \Delta u(x) = g(x, u(x))$$

$$u : \mathbb{R}^2 \to \mathbb{R}^n, \quad g : \mathbb{R}^2 \times \mathbb{R}^n \to \mathbb{R}^n, \quad n \geq 1$$

can also be treated in the spaces $E_q(\mathbb{R}^2 ; \mathbb{R}^n)$ and for some classes of examples the existence of a (weak) solution can be proven.

Acknowledgement:

I would like to thank Prof. G. Laßner for his kind invitation to Leipzig and for his attentive hospitality at the NTZ of the Karl-Marx-University of Leipzig where the original version of this note was written. With pleasure I also thank the staff of the NTZ for their many helps.

Finally I thank Proff. J. Ginibre and G. Velo for pointing out reference [9].

References:

1. Th. Aubin: Nonlinear analysis on manifolds. Monge-Ampère Equations. Springer Verlag Berlin Heidelberg New York 1982

2. H. Berestycki, Th. Gallouët, O. Kavian: Equations de champs scalaires euclidiens nonlinéaires dans le plan. C.R. Acad. Sc. Paris, t. 297 (Oct. 1983), Série I - 307

3. H. Brezis, E.H. Lieb: Minimum action solutions of vectorfield equations. Commun. Math. Phys. 96, 97 - 113 (1984)

3'.H. Brezis, E.H. Lieb: private communication

4. E. Brüning: On the variational approach to semilinear elliptic equations with scale-covariance. Preprint: IHES/P/86/53

5. Ph. Blanchard, E. Brüning: Variational Methods in Mathematical Physics. A unified approach. Springer Verlag Berlin Heidelberg New York to appear 1987/88; translated and extended edition of "Direkte Methoden der Variationsrechnung". Springer Verlag Wien New York 1982

6. D. Gilbarg, N.S. Trudinger: Elliptic partial differential equations of second order. Grundlehren der mathematischen Wissenschaften 224. Springer Verlag Berlin Heidelberg New York 1977

7. A. Kufner, O. John, S. Fučik: Function spaces. Noordhoff International Publishing Leyden 1977

8. J. Moser: A sharp form of an inequality of N. Trudinger. Ind. Univ. Math. J. 20 (1971), 1077 - 1092

9. L.R. Volevich, B.P. Paneyakh: Certain spaces of generalized functions and embedding theorems. Russian Mathematical Surveys 20 (1965), 1 - 73

SOME REMARKS ON THE NONLINEAR SCHRÖDINGER
EQUATION IN THE SUBCRITICAL CASE

Thierry Cazenave [1] and Fred B. Weissler [1][2]

[1] Analyse Numérique, Université Pierre et Marie Curie,

4, Place Jussieu, 75252 PARIS CEDEX 05, FRANCE.

[2] Department of Mathematics, Texas A&M University,

COLLEGE STATION, TX 77843-3368, USA

1 INTRODUCTION

We consider the Cauchy problem (initial value problem) for nonlinear Schrödinger equations in $\mathbf{R}^n$, of the form

$$iu_t + \Delta u = g(u) \quad , \quad u(0, \cdot) = \varphi(\cdot) \ . \tag{NLS}$$

Here u is a complex-valued function defined on $[0,T) \times \mathbf{R}^n$ for some T>0, φ is some initial condition defined on $\mathbf{R}^n$ and g is some nonlinear (local or non-local) mapping. In most of the examples that have been considered, g has some symmetry properties and is also the gradient of some functional G. Thus, at least formally, we have both conservation of charge and conservation of energy, that is

$$\int_{\mathbf{R}^n} |u(t,x)|^2 \, dx \ = \ \int_{\mathbf{R}^n} |\varphi(x)|^2 \, dx \ ,$$

$$\frac{1}{2} \int_{\mathbf{R}^n} |\nabla u(t,x)|^2 \, dx \ + \ G(u(t,\cdot)) \ = \ \frac{1}{2} \int_{\mathbf{R}^n} |\nabla \varphi(x)|^2 \, dx \ + \ G(\varphi(\cdot)) \ .$$

Clearly, the charge and energy involve the H^1-norm of the solution and therefore it is important to be able to solve the local Cauchy problem in the space $H^1(\mathbf{R}^n)$. Indeed, when this is possible, then global existence results follow easily from the above conservation laws and some conditions on G (for example G≥0). Obviously, in order to be able to do so, there are some necessary requirements on g; for example, g and G need to be well defined on H^1. In the applications, this will impose some "growth" conditions on g.

The initial value problem for the nonlinear Schrödinger equation in H^1 has been studied in the past few years, essentially by J Ginibre and G Velo [5,6,7,8] and by T Kato [9]. In the model case where $g(u)=|u|^{p-1}u$, the Cauchy problem is well posed in $H^1(\mathbf{R}^n)$ for $1 \leq p < (n+2)/(n-2)$. The methods are of a perturbative nature and rely basically on sharp dispersive properties of the linear equation. All the previous proofs require at some stage (for obtaining local estimates of the solution in $H^1(\mathbf{R}^n)$) differentiation of the equation with respect to x, and so they don't apply to nonlinearities for which the x-dependence is not smooth enough. For example, the results mentioned above do not cover the case where $g(u)=|u|^{p-1}u+Vu$, V being a non-smooth potential.

We present here a result that covers most of the previously known cases and that holds without any smoothness assumption on g(u) with respect to x. The proof proceeds by an approximation argument followed by a passage to the limit. Uniform estimates on the approximating solutions are obtained from the conservation of the energy, and the passage to the limit (as well as uniqueness) relies on the dispersive properties of the linear Schrödinger equation. Let us remark that we do not need the conservation of charge (g does not have to satisfy the corresponding symmetry properties) while we definitely need the conservation of the energy (g must be the gradient of some potential G). This is in contrast with the result of [9], which applies to local nonlinearities for which there is possibly no energy (but that are sufficiently smooth with respect to x).

In section 2 we state the main result and we give some examples of applications, and in section 3 we give a sketch of the proof. The reader is referred to [3] for the complete proof and to [2,4] for some related results in the critical case where the present method just fails.

2 THE MAIN RESULT

We begin by introducing some notation. We denote by H^k the Sobolev space $H^k(\mathbf{R}^n,\mathbf{C})$ for any integer k, equipped with its usual norm and scalar product (always considered as a real Hilbert or Banach space) and by L^p the space $L^p(\mathbf{R}^n,\mathbf{C})$ for any $p \in [1,\infty]$, also equipped with its usual norm. We denote by $\| \ \|_{H^k}$ (respectively $\| \ \|_{L^p}$) the norm in H^k (respectively L^p), and by $< \ , \ >$ the duality pairing between H^{-1} and H^1. p' is the conjugate exponent of p , given by $1/p + 1/p' = 1$. For a given $\varphi \in H^1$, we are interested in the initial value problem (NLS).

We now state the assumptions on the nonlinear interaction g. We assume that g is of the form $g = \sum_{k=1}^{N} g_k$, where $g_k \in C(H^1,H^{-1})$. For each of the g_k, we assume the following. There exists a function $C_k \in C(\mathbf{R}_+,\mathbf{R}_+)$, two numbers $r_k, \rho_\kappa \in [2,2n/(n-2))$ $(r_k, \rho_\kappa \in [2,\infty)$ if n=1,2), and a sequence $g_{k,m} \in C(L^2,L^2)$ such that

$$g_{k,m}(0)=0 \text{ and } g_{k,m} \text{ is Lipschitz continuous from bounded sets of } L^2 \text{ to } L^2. \qquad (1)$$

$$g_{k,m} \to g_k \text{ in } L^{(\rho k)'} \text{ as } m \to \infty, \text{ uniformly on bounded sets of } H^1. \qquad (2)$$

$$\text{There exists } G_{k,m} \in C^1(L^2,\mathbf{R}) \text{ such that } G_{k,m}(0)=0 \text{ and } g_{k,m}=(G_{k,m})'. \qquad (3)$$

$$\|g_{k,m}(v)-g_{k,m}(u)\|_{L^{(\rho k)'}} \le C_k(M)\|v-u\|_{L^{rk}} \text{ for } u,v \in H^1, \text{ with } \|u\|_{H^1} \le M \text{ and } \|v\|_{H^1} \le M. \qquad (4)$$

From (3) we have $G_{k,m}(u) = \int_0^1 <g_{k,m}(su),u> ds$, for every $u \in H^1$. Thus, if we set

$$G_k(u) = \int_0^1 <g_k(su),u> ds , \qquad (5)$$

then from (2), (5) and the embeddings $H^1 \subset L^{rk}$, $L^{(\rho k)'} \subset H^{-1}$, we get

$$G_{k,m} \to G_k \text{ as } m \to \infty, \text{ uniformly on bounded sets of } H^1. \qquad (6)$$

Finally, let us define the functionals G, G^m, E and E^m by

$$G^m(u) = \sum_{k=1}^{N} G_{k,m}(u) \quad \text{for } u \in H^1 ,\tag{7}$$

$$G(u) = \sum_{k=1}^{N} G_k(u) \quad \text{for } u \in H^1 ,\tag{8}$$

$$E^m(u) = \frac{1}{2} \int_{\mathbf{R}^n} |\nabla u(x)|^2 \, dx \; + \; G^m(u) \quad \text{for } u \in H^1 ,\tag{9}$$

$$E(u) = \frac{1}{2} \int_{\mathbf{R}^n} |\nabla u(x)|^2 \, dx \; + \; G(u) \quad \text{for } u \in H^1 .\tag{10}$$

We can now state our main result.

THEOREM 1. Assume that g satisfies the above hypotheses. Then for any $\varphi \in H^1$, there exists $T^*>0$ and a solution $u \in C([0,T^*),H^1) \cap C^1([0,T^*),H^{-1})$ of (NLS). In addition, we have the following properties.

(i) u is unique in $C([0,T'),H^1)$ for any $T'>0$,

(ii) either $T^*=\infty$ or else $T^*<\infty$ and $\|u(t)\|_{H^1} \to \infty$ as t $\uparrow T^*$,

(iii) $E(u(t)) = E(\varphi)$, for every $t \in [0,T^*)$.

Several remarks and comments are in order, concerning both the hypotheses on g and the statement of Theorem 1.

REMARK 1. It follows from conditions (1) and (2) that g(0)=0. We assume this only for the sake of simplicity. Allowing $g(0) \neq 0$ would result in adding a constant term $\varphi \in H^{-1}$ to the right hand side of (NLS), which would not be too difficult to handle.

REMARK 2. We assume that g is split into N terms g_k satisfying different conditions. This is rather natural since in the applications (see below) the nonlinearity can be the sum of several terms having properties that are actually different.

REMARK 3. When applying Theorem 1 to some particular example, what is given in general is g as the sum of several terms g_k . One would expect to need only assumptions on the g_k's. However, our assumptions ((1) to (4)) are on some approximating sequence $g_{k,m}$, which seems somewhat unnatural.

This phenomenon comes from the following technical difficulty. In the proof, we need to approximate g_k by some sequence $g_{k,m}$ satisfying (1) and (2). So if we assume only that g_k satisfies (4) and is the gradient of some functional $G_k \in C^1(H^1, R)$, we have to find some approximate sequence satisfying (1) and (2), and this is not obvious. For example if we think of $g_{k,m}(u) = \rho_m * g_k(\rho_m * u)$ where ρ_m is a sequence of mollifiers, then $g_{k,m}$ will satisfy (1) to (4) except that the convergence in (2) will not be uniform on bounded sets of H^1. However, let us point out that the approximating sequence is easily found in the important examples (see below).

REMARK 4. The solution of (NLS) does not need to satisfy the conservation of charge. However, if we assume in addition to the other hypotheses that $<g(u), iu> = 0$ for every $u \in H^1$, then we get conservation of charge. This is easily seen by multiplying the equation by iu, in the sense of the duality between H^{-1} and H^1.

REMARK 5. Some global existence results are easily obtained from property (ii) of Theorem 1, conservation of energy and conservation of charge (if any). See [5,6,7,8], [9] and the examples given below.

Let us now give some examples of nonlinearities that satisfy the hypotheses of Theorem 1.

EXAMPLE 1 (external potential). Let V be a real-valued function on R^n. Assume that $V \in L^\sigma + L^\infty$, with $\sigma \geq 1$, $\sigma > n/2$. Let g be given by $g(u) = Vu$ for $u \in H^1$ and let $V = V_1 + V_2$, with $V_1 \in L^\sigma$ and $V_2 \in L^\infty$. Then $g = g_1 + g_2$ with $g_1(u) = V_1 u$ and $g_2(u) = V_2 u$. Now choose $r_1 = p_1 = 2\sigma/(\sigma-1)$, $r_2 = p_2 = 2$, and set $V_{1,m}(x) = \text{Min}\{m, \text{Max}\{-m, V_1(x)\}\}$, $g_{1,m}(u) = V_{1,m} u$ and $g_{2,m} = g_2$. It is easily verified that g satisfies the hypotheses of Theorem 1 with

$$G_{k,m}(u) = \frac{1}{2} \int_{R^n} V_{k,m}(x) |u(x)|^2 \, dx \ .$$

In this case, we always have global existence and conservation of charge.

EXAMPLE 2 (Hartree-type nonlinearity, see also [7]). Let W be a real valued even function on R^n. Assume that $W \in L^\delta + L^\infty$ with $\delta \geq 1$, $\delta > n/4$. Let g be given by $g(u) = (W * |u|^2)u$ and let $W = W_1 + W_2$, with $W_1 \in L^\delta$ and $W_2 \in L^\infty$. Then $g = g_1 + g_2$ with $g_1(u) = (W_1 * |u|^2)u$ and $g_2(u) = (W_2 * |u|^2)u$. Now choose

$r_1=\rho_1=4\delta/(2\delta-1)$, $r_2=\rho_2=2$, and set $W_{1,m}(x)=\text{Min}\{m,\text{Max}\{-m,W_1(x)\}\}$, $g_{1,m}(u)=(W_{1,m}*|u|^2)u$ and $g_{2,m}=g_2$. Applying Young's inequality, it is easily seen that the hypotheses of Theorem 1 are fulfilled with

$$G_{k,m}(u) = \frac{1}{4} \int_{\mathbf{R}^n} (W_{k,m}*|u|^2) |u|^2 \, dx \; .$$

In this case we always have conservation of charge. All the solutiuons are global if, for example, the negative part of W belongs to $L^\nu+L^\infty$, with $\nu=1$ if $n=1$, $\nu>1$ if $n=2$ and $\nu=n/2$ if $n\geq3$.

EXAMPLE 3 (local nonlinearity, see also [5,6,8] and [9]). Let $f:\mathbf{R}^n\times\mathbf{C}\to\mathbf{C}$ be a measurable function. Assume that $f(x,0)=0$ almost everywhere and that there exists $M\geq0$ and $\alpha\in[0,4/(n-2))$ ($\alpha\in[0,\infty)$ if $n=1,2$) such that $|f(x,z_2)-f(x,z_1)| \leq M(1+|z_1|^\alpha+|z_2|^\alpha)|z_2-z_1|$ for almost all $x\in\mathbf{R}^n$ and all $z_1,z_2\in\mathbf{C}$. Assume that $f(x,z)=(z/|z|)f(x,|z|)$. Let g be given by $g(u)(x)=f(x,u(x))$. Then g satisfies the hypotheses of Theorem 1. Indeed, a family $g_{k,m}$ is easily found. For example, let f_1 and f_2 be given by $f_1(x,z)=f(x,z)$ if $|z|\leq1$, $f_1(x,z)=zf(x,1)$ if $|z|\geq1$, $f_2=f-f_1$. Let $f_{1,m}=f_1$ and let $f_{2,m}$ be defined by $f_{2,m}(x,z)=f_2(x,z)$ if $|z|\leq m$, $f_{2,m}(x,z)=(z/m)f_2(x,m)$ if $|z|\geq m$. Let the function $F_{k,m}$ be defined by

$$F_{k,m}(x,z) = \int_0^{|z|} f_{k,m}(x,s) \, ds \; .$$

Then a suitable sequence $g_{k,m}$ is given by $g_{k,m}(u)(x)=f_{k,m}(x,u(x))$ and

$$G_{k,m}(u) = \int_{\mathbf{R}^n} F_{k,m}(x,u(x)) \, dx \; .$$

Here also we have conservation of charge. The solutions are global for all the initial data if for example

$$\int_0^s f(x,\sigma) \, d\sigma \geq - C - C s^\delta \text{ for all } s\geq0 \text{ and for some } \delta\in[0,1+4/n) \; .$$

EXAMPLE 4. It is quite clear that if g^1, g^2,..., g^j satisfy the hypotheses of the Theorem, then so does $g=g^1+g^2+...+g^j$. Therefore Theorem 1 applies when the nonlinearity is any finite sum of the nonlinearities considered in the examples 1, 2 and 3.

3 SKETCH OF THE PROOF.

The dispersive properties of the Schrödinger equation that we need are described in the following Lemma.

LEMMA 1. Let $r,\rho \in [2,2n/(n-2))$ $(r,\rho \in [2,\infty]$ if $n=1$ and $r,\rho \in [2,\infty)$ if $n=2)$ and $q,\gamma \in (2,\infty]$ $(q,\gamma \in [2,\infty]$ if $n=1)$ with $2/q=n(1/2-1/r)$ and $2/\gamma=n(1/2-1/\rho)$. Let $T>0$, $u_0 \in L^2$ and $f \in L^{\gamma'}(0,T,L^{\rho'}(\mathbf{R}^n))$. Then there exists C depending only on n,r,ρ such that the solution u of

$$iu_t + \Delta u = f \ , \ u(0) = \varphi \ , \tag{LS}$$

satisfies

$$\|u\|_{L^q(0,T,L^r)} \leq C \ (\ \|f\|_{L^{\gamma'}(0,T,L^{\rho'})} + \|\varphi\|_{L^2}) \ . \tag{11}$$

Lemma 1 is proved in [8] for $f \equiv 0$ and in [10] (see also [9]) in the special cases $r=\rho$, $r=2$, $\rho=2$. The general case follows by interpolating between two of these three cases, depending on whether $r>\rho$ or $r<\rho$ (see [2] for a suitable interpolation theorem).

REMARK 6. It is immediate from Lemma 1 that if $f=f_1+...+f_M$, where each of the f_j satisfy the assumption of the lemma with exponents (γ_j,ρ_j), then

$$\|u\|_{L^q(0,T,L^r)} \leq C (\sum_{j=1}^{M} \|f_j\|_{L^{(\gamma j)'}(0,T,L^{(\rho j)'})} + \|\varphi\|_{L^2}) \quad .$$

Now we can proceed to prove Theorem 1. We assume that g satisfies the hypotheses and we consider $\varphi \in H^1$. For $m \in N$ we consider the solution u^m of the problem

$$i(u^m)_t + \Delta u^m = g^m(u^m) \ , \ u^m(0,\cdot) = \varphi(\cdot) \ , \tag{NLS,m}$$

where $g^m = g_{1,m} + ... + g_{N,m}$. It is not too difficult to prove (see [3]) that

$u^m \in C([0,T^m),H^1) \cap C^1([0,T^m),H^{-1})$ for some $T^m > 0$ and that we have

$$E^m(u^m(t)) = E^m(\varphi) \text{, for every } t \in [0,T^m). \tag{12}$$

The next step is the following.

LEMMA 2. There exists $T_1 > 0$ depending on $\|\varphi\|_{H^1}$ such that $\|u^m\|_{L^\infty(0,T1,H^1)} \leq 2 \|\varphi\|_{H^1}$.

PROOF. Let $[0,T_m]$ be the maximal interval on which $\|u^m(\cdot)\|_{H^1} \leq 2 \|\varphi\|_{H^1}$. All we need is a positive lower bound on T_m. Now it follows from (4), (1) and the equation that $\|(u^m)_t\|_{H^{-1}}$ is bounded on $[0,T_m]$ by some K independent of m. Therefore, there exists a constant K' such that

$$\|u^m(t)-\varphi\|_{L^2} \leq K' \, t^{1/2} \text{, for } t \in [0,T_m] . \tag{13}$$

From (12) we obtain

$$(\|u^m(t)\|_{H^1})^2 = (\|\varphi\|_{H^1})^2 + G^m(\varphi) - G^m(u^m(t)) + (\|u^m(t)\|_{L^2})^2 - (\|\varphi\|_{L^2})^2 . \tag{14}$$

In (14) now, we estimate the L^2 terms using (13) and the G^m terms using (4). Together with Gagliardo-Nirenberg and Sobolev's inequalities, we get for some K" independent of m, some $\delta > 0$, and for, say, $t \leq 1$,

$$(\|u^m(t)\|_{H^1})^2 \leq (\|\varphi\|_{H^1})^2 + K'' \, t^\delta . \tag{15}$$

(See [3] for the details of this calculation.) For t less than some $T_0 > 0$, the right-hand side of (15) is less than $4(\|\varphi\|_{H^1})^2$, and so T_m is bounded from below by T_0 .

The next Lemma is crucial for the passage to the limit.

LEMMA 3. u^m is a Cauchy sequence in $C([0,T_2],L^2)$ for some $T_2 > 0$ depending on $\|\varphi\|_{H^1}$.

PROOF. Let $j,m \in \mathbf{N}$. From (NLS,j) and (NLS,m) we obtain $(u^j - u^m)(0) = 0$ and

$$i\,(u^j - u^m)_t + \Delta\,(u^j - u^m) = \sum_{k=1}^{N} (g_{k,j}(u^j) - g_{k,j}(u^m)) + \sum_{k=1}^{N} (g_{k,j}(u^m) - g_{k,m}(u^m)) \ .$$

Let $r \in [2, 2n/(n-2))$ ($r \in [2, \infty)$ if $n=1,2$) and $q \in (2, \infty]$ with $2/q = n(1/2 - 1/r)$. We apply Lemma 1 and Remark 6 to estimate $(u^j - u^m)$ in $L^q(0,T,L^r(\mathbf{R}^n))$, for some $T \le T_1$. To this end we introduce the exponents γ_k given by $2/\gamma_k = n(1/2 - 1/\rho_k)$, and we estimate the terms $(g_{k,j}(u^j) - g_{k,j}(u^m))$ and $(g_{k,j}(u^m) - g_{k,m}(u^m))$ in $L^{(\gamma k)'}(0,T,L^{(\rho k)'}(\mathbf{R}^n))$. By Lemma 2 and (2), we obtain

$$\sum_{k=1}^{N} \|g_{k,j}(u^m) - g_{k,m}(u^m)\|_{L^{(\gamma k)'}(0,T,L^{(\rho k)'})} \to 0 \quad \text{as } j,m \to \infty \ .$$

We estimate the terms $(g_{k,j}(u^j) - g_{k,j}(u^m))$ by using (4) together with Hölder's inequality on $(0,T)$. This yields

$$\sum_{k=1}^{N} \|g_{k,j}(u^j) - g_{k,j}(u^m)\|_{L^{(\gamma k)'}(0,T,L^{(\rho k)'})} \le C \sum_{k=1}^{N} T^{\alpha k} \|u^j - u^m\|_{L^{qk}(0,T,L^{rk})} \ ,$$

where C is some constant depending on $\|\varphi\|_{H^1}$, q_k is given by $2/q_k = n(1/2 - 1/r_k)$, and $\alpha_k = (q_k - 2)/q_k > 0$. The choice of (q,r) is arbitrary, so we choose successively $(q,r) = (\infty, 2)$, and $(q,r) = (q_k, r_k)$ for $k = 1, \ldots, N$. Adding the resulting inequalities, and choosing T small enough (depending on $\|\varphi\|_{H^1}$), we find that

$$M(T) \le \varepsilon(j,m) + (1/2)\,M(T) \ ,$$

where $\varepsilon(j,m) \to 0$ as $j,m \to \infty$ and

$$M(T) = \|u^j - u^m\|_{L^\infty(0,T,L^2)} + \sum_{k=1}^{N} \|u^j - u^m\|_{L^{qk}(0,T,L^{rk})} \ .$$

Hence the result.

We are now in position to complete the proof of Theorem 1. First, uniqueness is obtained with the same technique as in the proof of Lemma 3 since the g_k's satisfy the same estimates as the $g_{k,m}$'s do. Now we consider the sequence u^m defined above, and we let $T=Min(T_1,T_2)$ where T_1 and T_2 are given by Lemmas 2 and 3. Let u be the limit of u^m in $C(0,T,L^2)$. From the uniform H^1 bound on the u^m, we get also $u \in L^\infty(0,T,H^1)$ and by Sobolev's inequality, $u^m \to u$ in $C(0,T,L^r)$ for any $r \in [2,2n/(n-2))$ ($r \in [2,\infty)$ if n=1,2). Therefore, it follows from (2) and (4) that $g^m(u^m) \to g(u)$ in $C(0,T,H^{-1})$. Thus u solves (NLS) in $L^\infty(0,T,H^{-1})$. Now, from (4), (6), (12) and the weak lower semicontinuity of the H^1-norm we obtain that $E(u(t)) \le E(\varphi)$ for $t \in [0,T]$. Reversing the sense of time and using uniqueness we obtain the same property for $w(s)=u(t-s)$, $s \in [0,t]$, and in particular we get conservation of the energy (property (iii)). Therefore, the map $t \to \|u(t)\|_{H^1}$ is continuous. Since u is weakly continuous in H^1, it follows that in fact $u \in C(0,T,H^1)$ and then also that $u \in C^1(0,T,H^{-1})$.

Thus we have established the existence of the solution described in Theorem 1 (all properties except (ii)) on an interval [0,T] where T depends only on $\|\varphi\|_{H^1}$. We now extend u to be a maximal solution on [0,T*) and property (ii) follows easily. This proves Theorem 1.

REFERENCES

[1] J BERGH, J LÖFSTRÖM. Interpolation spaces. Springer, New York, 1976.

[2] T CAZENAVE, F B WEISSLER. Some remarks on the nonlinear Schrödinger equation in the critical case. Proceedings of the Second Howard University Symposium on Nonlinear Semigroups, Partial Differential Equations, and Attractors. Washington, D.C., August 1987. Springer, to appear.

[3] T CAZENAVE, F B WEISSLER. The Cauchy problem for the nonlinear Schrödinger equation in H^1. To appear.

[4] T CAZENAVE, F B WEISSLER. The Cauchy problem for the critical nonlinear Schrödinger equation in H^s. To appear.

[5] J GINIBRE, G VELO. On a class of nonlinear Schrödinger equations. J. Funct. Anal., 32 (1979), 1-71.

[6] J GINIBRE, G VELO. On a class of nonlinear Schrödinger equations. Special theories in dimensions 1, 2 and 3. Ann. Inst. Henri Poincaré, **28** (1978), 287-316.

[7] J GINIBRE, G VELO. On a class of nonlinear Schrödinger equations with non local interaction. Math. Z., **170** (1980), 109-136.

[8] J GINIBRE, G VELO. The global Cauchy problem for the nonlinear Schrödinger equation revisited. Ann. Inst. Henri Poincaré, Analyse Non Linéaire, **2** (1985), 309-327.

[9] T KATO. On nonlinear Schrödinger equations. Ann. Inst. Henri Poincaré, Physique Théorique, **46** (1987), 113-129.

[10] K YAJIMA. Existence of solutions for Schrödinger evolution equations. Comm. Math. Phys., **110** (1987), 415-426.

The Cauchy problem for the Dirac equation with cubic nonlinearity in three space dimensions

João-Paulo Dias and Mário Figueira

CMAF

2, Av. Prof. Gama Pinto

P-1699 Lisboa Codex

1. Introduction

Let us consider the nonlinear Dirac equation in $\mathbb{R} \times \mathbb{R}^3$

$$(1.1) \quad i \frac{\partial \psi}{\partial t} = -i\alpha.\nabla\psi + m\beta\psi + k(\psi^+\beta\psi)\beta\psi,$$

$$\psi = \psi(t,x), \quad m \geq 0, \quad k \in \mathbb{R},$$

where $\alpha.\nabla\psi = \sum\limits_{j=1}^{3} \alpha_j \frac{\partial \psi}{\partial x_j}$,

$$\alpha_1 = \begin{bmatrix} 0 & 0 & 0 & 1 \\ 0 & 0 & 1 & 0 \\ 0 & 1 & 0 & 0 \\ 1 & 0 & 0 & 0 \end{bmatrix}, \quad \alpha_2 = \begin{bmatrix} 0 & 0 & 0 & -i \\ 0 & 0 & i & 0 \\ 0 & -i & 0 & 0 \\ i & 0 & 0 & 0 \end{bmatrix}, \quad \alpha_3 = \begin{bmatrix} 0 & 0 & 1 & 0 \\ 0 & 0 & 0 & -1 \\ 1 & 0 & 0 & 0 \\ 0 & -1 & 0 & 0 \end{bmatrix}, \quad \alpha_4 = \beta = \begin{bmatrix} 1 & 0 & 0 & 0 \\ 0 & 1 & 0 & 0 \\ 0 & 0 & -1 & 0 \\ 0 & 0 & 0 & -1 \end{bmatrix}$$

$(\alpha_j^\dagger = \bar{\alpha}_j^T = \alpha_j, \; \alpha_j^2 = I, \; \alpha_j \alpha_m = -\alpha_m \alpha_j$ for $j \neq m)$

and $\psi: \mathbb{R} \times \mathbb{R}^3 \to \mathbb{C}^4$ is a column vector $(\psi_1, \psi_2, \psi_3, \psi_4)$, $\psi^+ = \bar{\psi}^T$. We put $|\psi|^2 = \psi^+\psi$. The operator $A = -i\alpha.\nabla + m\beta$ is self-adjoint in $\mathbb{L}^2 = (L^2(\mathbb{R}^3))^4$ with domain $\mathbb{H}^1$. We can consider the unitary group $S(t) = e^{-itA} (t \in \mathbb{R})$ in $\mathbb{H}^s$, $s \geq 0$, and so we write the Cauchy problem for the equation (1.1) in the following integral form

$$(1.2) \quad \psi(t) = S(t)\psi_0 - i\int_0^t S(t-\tau)J(\psi(\tau)) \, d\tau,$$

where $\psi_0 \in \mathbb{H}^s$, $s \geq 2$, and $J\psi = k(\psi^+\beta\psi)\beta\psi$. The function $w(t,x) = S(t)\psi_0$ veryfies the Klein-Gordon equation (wave equation if m=0) with

$$\frac{\partial}{\partial t}\left[S(t)\psi_0\right]_{t=0} = -\alpha.\nabla\psi_0 - im\beta\psi_0.$$

We recall that a solution $\psi \in C(\mathbb{R}; \mathbb{H}^1)$ of (1.1) satisfies the conservation laws

$$(1.3) \quad |\psi(t)|_2 = |\psi_0|_2, \quad t \in \mathbb{R}.$$

$$(1.4) \quad \text{Im} \int \psi^+(t)\alpha.\nabla\psi(t)dx + m\int \psi^+(t)\beta\psi(t)dx +$$

$$+ \frac{k}{2} \int (\psi^+(t)\beta\psi(t))^2 \, dx = \text{(energy)}, \quad t \in \mathbb{R},$$

where $\int = \int_{\mathbb{R}^3}$. The energy will not be useful in our estimates. A local existence and uniqueness theorem for the Cauchy problem for equation (1.1) is easy to prove if $\psi_0 \in \mathbb{H}^2$. In [7], M. Reed has obtained a global existence result for equation (1.1) with higher nonlinearity, if $m > 0$, $\psi_0 \in \mathbb{H}^3$ and $\|\psi_0\| = |\psi_0|_{\mathbb{H}^3} + \sup_{t \in \mathbb{R}}[(1 + |t|)^{3/2} |S(t)\psi_0|_\infty]$ is small enough (cf. [7], theorem 2.2). The cubic case (if $m > 0$) can also be included in his result (cf. [3]). Furthermore, if $m > 0$ and $\psi_0 \in \mathbb{H}^{2+s}$, $0 < s < 1$, we can modify the work of M. Reed in order to obtain an existence result of the same kind, with $\frac{3}{2}$ replaced by $(1 + \frac{1}{2} s)(1 - \varepsilon)$, $\varepsilon > 0$ such that $(1 + \frac{1}{2} s)(1 - \varepsilon)^2 > 1$(cf. [3]). In §3 of this paper we extend this result to $\psi_0 \in \mathbb{H}^2$, replacing the $\mathbb{L}^\infty$ norm by an $\mathbb{W}^{1,p}$ norm, $3 < p \leq \frac{10}{3}$, and applying the estimates of Ph. Brenner for the linear Klein-Gordon equation (cf. [2]). In §2 we study the null mass case ($m = 0$). In this case we need $\psi_0 \in \mathbb{H}^3$ but, as J. Ginibre and G. Velo pointed out to us, this condition can be weakned in the framework of Besov spaces. Then, replacing the $\mathbb{L}^\infty$ norm by an $\mathbb{W}^{1,p}$ norm, $4 < p < +\infty$, and applying the estimates of H. Pecher for the linear wave equation (cf. [5] and [6]) we obtain a global existence result for the Cauchy problem if

$$\|\psi\|_\Sigma = |\psi_0|_{\mathbb{H}^3} + \sup_{t \in \mathbb{R}} [(1 + |t|)^{1 - \frac{2}{p}} |S(t)\psi_0|_{\mathbb{W}^{1,p}}]$$

is small enough (cf. [4]).

2. The null mass case ($m = 0$).

Assume $m = 0$, $\phi \in \mathbb{H}^3$. Let $\mathring{H}^{s,p'}(\mathbb{R}^3)$, $s \in \mathbb{R}$, $4 < p < +\infty$, $\frac{1}{p} + \frac{1}{p'} = 1$, be the completion of $\mathcal{D}(\mathbb{R}^3)$ with respect to $|F^{-1}(|\xi|^s \hat{f}(\xi))|_p$, $f \in \mathcal{D}(\mathbb{R}^3)$. If $\phi \in \mathbb{\mathring{H}}^{s,p'} = (\mathring{H}^{s,p'}(\mathbb{R}^3))^4$ with $s = 2 - \frac{4}{p}$, we have (cf. [5] and [6]):

$$|S(t)\phi|_p \leq c |t|^{-1 + \frac{2}{p}} (|\phi|_{\mathring{H}^{s,p'}} + |\alpha \cdot \nabla\phi|_{\mathring{H}^{s-1,p'}}), \quad t \neq 0.$$

Since we have (cf. [1]) $W^{2,p'} = H^{2,p'} \hookrightarrow H^{s,p'} \hookrightarrow \mathring{H}^{s,p'}$, where $H^{s,p'}$ is the completion of $\mathcal{D}(\mathbb{R}^3)$ with respect to $|F^{-1}[(1+|\xi|^2)^{s/2} \hat{f}(\xi)]|_{p'}$, we obtain

(2.1) $\quad |S(t)\phi|_{\mathbb{W}^{1,p}} \leq c|t|^{-1 + \frac{2}{p}} |\phi|_{\mathbb{W}^{3,p'}}$, if $\phi \in \mathbb{W}^{3,p'}$, $t \neq 0$.

Furthermore, since $|S(t)\phi|_{\mathbb{W}^{1,p}} \leq c|S(t)\phi|_{\mathbb{H}^3} = c|\phi|_{\mathbb{H}^3}$, we have

(2.2) $\quad |S(t)\phi|_{\mathbb{W}^{1,p}} \leq c |\phi|_{\mathbb{H}^3}$, $t \in \mathbb{R}$.

Hence, by (2.1) and (2.2), we obtain

$$(2.3) \quad |S(t)\phi|_{\mathbb{W}^{1,p}} \le c(1+|t|)^{-1+\frac{2}{p}}(|\phi|_{\mathbb{H}^3} + |\phi|_{\mathbb{W}^{3,p'}}),$$

if $\phi \in \mathbb{H}^3 \cap \mathbb{W}^{3,p'}$ and $t \in \mathbb{R}$.

We need the following result:

<u>Lemma 2.1</u>: Assume $\psi_1, \psi_2 \in \mathbb{H}^3$, $J\psi = k(\psi^+\beta\psi)\beta\psi$. We have

$$(2.4) \quad |J\psi_1 - J\psi_2|_{\mathbb{H}^3} \le c\,\gamma(\psi_1,\psi_2), \text{ if } 4<p<+\infty,$$

with $\gamma(\psi_1,\psi_2) = [(|\psi_1|^2_{\mathbb{W}^{1,p}} + |\psi_2|^2_{\mathbb{W}^{1,p}})|\psi_1-\psi_2|_{\mathbb{H}^3} +$

$$+ (|\psi_1|_{\mathbb{H}^3} + |\psi_2|_{\mathbb{H}^3})(|\psi_1|_{\mathbb{W}^{1,p}} + |\psi_2|_{\mathbb{W}^{1,p}})|\psi_1-\psi_2|_{\mathbb{W}^{1,p}}].$$

$$(2.5) \quad |J\psi_1 - J\psi_2|_{\mathbb{W}^{3,p'}} \le c\gamma(\psi_1,\psi_2) \text{ if } 4<p\le6.$$

$$(2.6) \quad |J\psi_1 - J\psi_2|_{\mathbb{W}^{3,p'}} \le$$

$$\le c[(|\psi_1|_{\mathbb{W}^{1,p}} + |\psi_2|_{\mathbb{W}^{1,p}})(|\psi_1|_{\mathbb{W}^{1,r}} + |\psi_2|_{\mathbb{W}^{1,r}})|\psi_1-\psi_2|_{\mathbb{H}^3} +$$

$$+ (|\psi_1|_{\mathbb{H}^3} + |\psi_2|_{\mathbb{H}^3})(|\psi_1|_{\mathbb{W}^{1,r}} + |\psi_2|_{\mathbb{W}^{1,r}})|\psi_1-\psi_2|_{\mathbb{W}^{1,p}}],$$

if $6<p<+\infty$, with $\frac{1}{r} = \frac{1}{2} - \frac{2}{p}$.

<u>Proof</u>: Let us take the "simplified model" $Ju = u^3$, $u \in H^3(\mathbb{R}^3)$ real and let us estimate $D^3(u^3 - v^3)$, where D^3 is a third derivative. Let $\omega = u-v$. The terms of $D^3(u^3 - v^3)$ are of the type $\theta_1 = D^3 \omega\, u\, v$, $\theta_2 = D^2\omega\, Du\, v$, $\theta_3 = D\omega\, D^2u\, v$, $\theta_4 = D\omega\, Du\, Dv$, $\theta_5 = \omega\, D^3u\, v$, $\theta_6 = \omega\, D^2u\, Dv$. Let $4<p<+\infty$. We have, by Holder's inequality and Sobolev's imbedding theorem:

$$|\theta_1|_2 \le |D^3\omega|_2\, |u|_\infty\, |v|_\infty \le c\,\gamma_1(\omega,u,v),$$

$$|\theta_2|_2 \le |D^2\omega|_s\, |Du|_p\, |v|_\infty \le c\,\gamma_1(\omega,u,v),$$

with $\gamma_1(\omega,u,v) = |\omega|_{H^3}|u|_{W^{1,p}}|v|_{W^{1,p}}$ and $\frac{1}{s} = \frac{1}{2} - \frac{1}{p}$,

$$|\theta_3|_2 \le |D\omega|_p\, |D^2u|_s\, |v|_\infty \le c\,\gamma_1(u,\omega,v),$$

$$|\theta_4|_2 \le |D\omega|_p\, |Du|_r\, |Dv|_p \le c\,\gamma_1(u,\omega,v),$$

$$|\theta_5|_2 \le |\omega|_\infty|D^3u|_2\, |v|_\infty \le c\,\gamma_1(u,\omega,v),$$

$$|\theta_6|_2 \le |\omega|_\infty |D^2 u|_s |Dv|_p \le c\ \gamma_1(u,\omega,v),$$

with $\frac{1}{s} = \frac{1}{2} - \frac{1}{p}$ and $\frac{1}{r} = \frac{1}{2} - \frac{2}{p}$.

Now, let $4 < p \le 6$, $\frac{1}{p'} + \frac{1}{p} = 1$. We have

$$|\theta_1|_{p'} \le |D^3\omega|_2 |u|_q |v|_q \le c\gamma_1(\omega,u,v),$$

$$|\theta_2|_{p'} \le |D^2\omega|_2 |Du|_p |v|_r \le c\ \gamma_1(\omega,u,v),$$

$$|\theta_3|_{p'} \le |D\omega|_p |D^2u|_2 |v|_r \le c\ \gamma_1(u,\omega,v),$$

$$|\theta_4|_{p'} \le |D\omega|_p |Du|_p |Dv|_{r_1} \le c\gamma_1(v,\omega,u),$$

$$|\theta_5|_{p'} \le |\omega|_r |D^3u|_2 |v|_p \le c\ \gamma_1(u,\omega,v),$$

$$|\theta_6|_{p'} \le |\omega|_r |D^2u|_2 |Dv|_p \le c\ \gamma_1(u,\omega,v),$$

with $\frac{1}{q} = \frac{1}{4} - \frac{1}{2p}$; $\frac{1}{r} = \frac{1}{2} - \frac{2}{p}$ and $\frac{1}{r_1} = 1 - \frac{3}{p}$.

Finally, let $6 < p < +\infty$. We have,

$$|\theta_1|_{p'} \le |D^3\omega|_2 |u|_p |v|_r \le c\ \gamma_2(\omega,u,v),$$

$$|\theta_2|_{p'} \le |D^2\omega|_2 |Du|_p |v|_r \le c\ \gamma_2(\omega,u,v),$$

$$|\theta_3|_{p'} \le |D\omega|_p |D^2u|_2 |v|_r \le c\ \gamma_2(\omega,u,v),$$

$$|\theta_4|_{p'} \le |D\omega|_p |Du|_2 |Dv|_r \le c\gamma_2(u,\omega,v),$$

$$|\theta_5|_{p'} \le |\omega|_p |D^3u|_2 |v|_r \le c\ \gamma_2(u,\omega,v),$$

$$|\theta_6|_{p'} \le |\omega|_p |D^2u|_2 |Dv|_r \le c\ \gamma_2(u,\omega,v),$$

with $\gamma_2(\omega,u,v) = |\omega|_{H^3} |u|_{W^{1,p}} |v|_{W^{1,r}}$ and $\frac{1}{r} = \frac{1}{2} - \frac{2}{p}$.

Notice that, if $6 < p < +\infty$, we have $6 > r > 2$ and $(1 - \frac{2}{p}) + (1 - \frac{2}{r}) = 1 + \frac{2}{p} > 1$. ▨

Now, let $4 < p < +\infty$, $\psi \in C(\mathbb{R}\ ;\ \mathbb{H}^3)$. We put

$$(2.7)\quad \| \psi \| = \sup_{t \in \mathbb{R}} |\psi(t)|_{\mathbb{H}^3} + \sup_{t \in \mathbb{R}} \left[(1 + |t|)^{1 - \frac{2}{p}} |\psi(t)|_{W^{1,p}} \right]$$

Let $\Sigma = \{\phi \in \mathbb{H}^3 \mid \|\phi\|_\Sigma < +\infty\}$, where

(2.8) $\|\phi\|_{\Sigma} = \|S(t)\phi\| = |\phi|_{{I\!H}^3} + \sup_{t\in{I\!R}}\left[(1+|t|)^{1-\frac{2}{p}} |S(t)\phi|_{{I\!W}^{1,p}}\right]$

By (2.3), if $\phi\in {I\!H}^3\cap {I\!W}^{3,p'}$, $\frac{1}{p'}+\frac{1}{p}=1$, we have $\phi\in\Sigma$ and

$$\|\phi\|_{\Sigma} \leq (|\phi|_{{I\!H}^3} + |\phi|_{{I\!W}^{3,p}}).$$

We can now prove the following theorem (cf. [4]):

Theorem 2.1: Assume $\psi_0\in {I\!H}^3$ and suppose that, for $p\in]4,+\infty[$, $\|\psi_0\|_{\Sigma}$ is small enough. Then, there exists an unique solution ψ of the equation (1.1) (m=0) verifying $\psi(0)=\psi_0$ and such that $\psi\in C({I\!R}; {I\!H}^3)\cap C^1({I\!R}; {I\!H}^2)$. Furthermore, $\|\psi\| <+\infty$.

Proof: Let $\eta>0$ and $\psi_0\in\Sigma$ such that $\|\psi\|_{\Sigma}\leq \eta$. We put

$$X(\eta,\psi_0) = \{\psi\in C({I\!R}; {I\!H}^3)|\ \|\psi-S(t)\psi_0\| \leq \eta\},$$

which is a complete metric space for the distance $d(\psi_1,\psi_2) = \|\psi_1-\psi_2\|$. We have $\|\psi\|\leq 2\eta$ for $\psi\in X(\eta,\psi_0)$. Define $(T\psi)(t) = -i\int_0^t S(t-\tau)J(\psi(\tau))d\tau$, $(M\psi)(t) = S(t)\psi_0 + (T\psi)(t)$, for $\psi\in X(\eta,\psi_0)$, with $J\psi = k(\psi^+\beta\psi)\beta\psi$. We assume that $4<p\leq 6$, since the proof in the case $6<p<+\infty$ follows by an easy adaptation (cf. lemma 2.1). By reasons of symmetry we take $t\geq 0$. We have, by (2.4) and since $2-\frac{4}{p}>1$,

$$|(T\psi)(t)|_{{I\!H}^3} \leq \int_0^t |J(\psi(\tau))|_{{I\!H}^3}\ d\tau \leq$$

$$\leq c_2(2\eta)^3\int_0^t (1+|\tau|)^{-2+\frac{4}{p}} d\tau \leq c_3\eta^3 .$$

Now, by (2.3), (2.4) and (2.5), we deduce

$$|(T\psi)(t)|_{{I\!W}^{1,p}} \leq \int_0^t |S(t-\tau) J(\psi(\tau))|_{{I\!W}^{1,p}} d\tau \leq$$

$$\leq c\int_0^t (1+|t-\tau|)^{-1+\frac{2}{p}} (|J(\psi(\tau))|_{{I\!H}^3} + |J(\psi(\tau))|_{{I\!W}^{3,p'}})d\tau$$

$$\leq c_4\int_0^t (1+|t-\tau|)^{-1+\frac{2}{p}}|\psi(\tau)|_{{I\!H}^3}|\psi(\tau)|^2_{{I\!W}^{1,p}}\ d\tau \leq$$

$$\leq c_4(2\eta)^3 \int_0^t (1+|t-\tau|)^{-1+\frac{2}{p}} (1+|\tau|)^{-2+\frac{4}{p}} d\tau \leq$$

$$\leq c_5\ \eta^3(1+|t|)^{-1+\frac{2}{p}} ,\ \text{by the lemma in page 78 of [7]. Hence,}$$

$\|T\psi\| \leq c_6\ \eta^3$ if $\psi\in X(\eta,\psi_0)$ and so $\|M\psi-S(t)\psi_0\| \leq c_6\ \eta^3$. Let us take $\eta_0>0$ such that $c_6\eta_0^2\leq 1$. Hence, for $\eta\leq\eta_0$, we have proved that

$M : X(n, \psi_0) \to X(n, \psi_0)$. By a similar argument we can prove that, if ψ_1, $\psi_2 \in X(n, \psi_0)$ with $n \leq n_0$, we have

$$|(M\psi_1)(t) - (M\psi_2)(t)|_{\mathbb{H}^3} \leq c_8 \, n^2 \| \psi_1 - \psi_2 \|,$$

$$|(M\psi_1)(t) - (M\psi_2)(t)|_{W^{1,p}} \leq c_9 \, n^2 \| \psi_1 - \psi_2 \| \, (1 + |t|)^{-1 + \frac{2}{p}}.$$

Hence $\| M\psi_1 - M\psi_2 \| \leq c_{10} \, n^2 \| \psi_1 - \psi_2 \|$. By choosing $n_1 \leq n_0$ such that $c_{10} \, n_1^2 < 1$, we can apply Banach's fixed point theorem and this completes the proof. ▨

3. The massive case (m>0).

Assume $m > 0$, $\phi \in \mathbb{H}^2$, $3 < p \leq \frac{10}{3}$ (hence, $3\left(\frac{1}{2} - \frac{1}{p}\right) > \frac{1}{2}$ and $\frac{3}{2} - \frac{5}{p} \leq 0$). We have (cf. [2] and [6]) if $\phi \in \mathbb{H}^{s,p'}$, with $s = 1 + \frac{3}{2} - \frac{5}{p}$, $\frac{1}{p'} + \frac{1}{p} = 1$,

$$|S(t)\phi|_p \leq c|t|^{-3\left(\frac{1}{2} - \frac{1}{p}\right)} \left(|\phi|_{\mathbb{H}^{s,p'}} + |\alpha.\nabla\phi|_{\mathbb{H}^{s-1,p'}} \right), \quad \text{for } |t| \geq 1.$$

Hence, since $W^{1,p'} = H^{1,p'} \hookrightarrow H^{s,p'}$, we have

$$|S(t)\phi|_p \leq c \, |t|^{-3\left(\frac{1}{2} - \frac{1}{p}\right)} |\phi|_{W^{1,p'}}, \quad |t| \geq 1,$$

if $\phi \in W^{1,p'}$, and so

$$(3.1) \quad |S(t)\phi|_{W^{1,p}} \leq c \, |t|^{-3\left(\frac{1}{2} - \frac{1}{p}\right)} |\phi|_{W^{2,p'}}, \quad |t| \geq 1,$$

if $\phi \in W^{2,p'}$.

Furthermore, since $|S(t)\phi|_{W^{1,p}} \leq c |S(t)\phi|_{\mathbb{H}^2} = c |\phi|_{\mathbb{H}^2}$, we have

$$(3.2) \quad |S(t)\phi|_{W^{1,p}} \leq c |\phi|_{\mathbb{H}^2}, \quad t \in \mathbb{R}.$$

Hence, by (3.1) and (3.2), we obtain

$$(3.3) \quad |S(t)\phi|_{W^{1,p}} \leq c(1 + |t|)^{-3\left(\frac{1}{2} - \frac{1}{p}\right)} \left(|\phi|_{\mathbb{H}^2} + |\phi|_{W^{2,p'}} \right),$$

if $\phi \in \mathbb{H}^2 \cap W^{2,p'}$ and $t \in \mathbb{R}$.

We need the following result:

Lemma 3.1: Assume $\psi_1, \psi_2 \in \mathbb{H}^2$, $J\psi = k(\psi^{+}\beta\psi)\beta\psi$. We have

$$(3.4) \quad \max\left(|J\psi_1 - J\psi_2|_{\mathbb{H}^2}, \ |J\psi_1 - J\psi_2|_{W^{2,p'}} \right) \leq$$

$$\leq c\left[\left(|\psi_1|^2_{W^{1,p}} + |\psi_2|^2_{W^{1,p}}\right)|\psi_1 - \psi_2|_{H^2} + \right.$$

$$\left. + \left(|\psi_1|_{H^2} + |\psi_2|_{H^2}\right)\left(|\psi_1|_{W^{1,p}} + |\psi_2|_{W^{1,p}}\right)|\psi_1 - \psi_2|_{W^{1,p}}\right].$$

Proof: Let us take again the "simplified model" $Ju = u^3$, $u \in H^3(\mathbb{R}^3)$ real and let us estimate $D^2(u^3 - v^3)$ where D^2 is a second derivative. Let $\omega = u - v$. The terms of $D^2(u^3 - v^3)$ are of the type $\theta_1 = D^2\omega\, u\, v$, $\theta_2 = D\omega D u v$, $\theta_3 = \omega D^2 u\, v$. Let $3 < p \leq \frac{10}{3}$. We have, by Holder's inequality and Sobolev's imbedding theorem:

$$|\theta_1|_2 \leq |D^2\omega|_2 \ |u|_\infty \ |v|_\infty \leq c\,\gamma_3(\omega, u, v),$$

$$|\theta_2|_2 \leq |D\omega|_s \ |Du|_p \ |v|_\infty \leq c\,\gamma_3(\omega, u, v),$$

$$|\theta_3|_2 \leq |\omega|_\infty \ |D^2 u|_2 \ |v|_\infty \leq c\,\gamma_3(u, \omega, v),$$

with $\gamma_3(\omega, u, v) = |\omega|_{H^2} \ |u|_{W^{1,p}} \ |v|_{W^{1,p}}$ and $\frac{1}{s} = \frac{1}{2} - \frac{1}{p}$.
Furthermore, we have, with $\frac{1}{p'} + \frac{1}{p} = 1$,

$$|\theta_1|_{p'} \leq |D^2\omega|_2 \ |u|_q \ |v|_q \leq c\,\gamma_3(\omega, u, v),$$

$$|\theta_2|_{p'} \leq |D\omega|_p \ |Du|_p \ |v|_{r_1} \leq c\,\gamma_3(v, \omega, u),$$

$$|\theta_2|_{p'} \leq |\omega|_q \ |D^2 u|_2 \ |v|_q \leq c\,\gamma_3(u, \omega, v),$$

with $\frac{1}{q} = \frac{1}{4} - \frac{1}{2p}$ and $\frac{1}{r_1} = 1 - \frac{3}{p}$.

Now, let $3 < p \leq \frac{10}{3}$, $\psi \in C(\mathbb{R}; H^2)$. We put

$$(3.8) \quad |||\psi||| = \sup_{t \in \mathbb{R}} |\psi(t)|_{H^2} + \sup_{t \in \mathbb{R}} \left[(1 + |t|)^{3\left(\frac{1}{2} - \frac{1}{p}\right)} |\psi(t)|_{W^{1,p}}\right]$$

Let $\tilde{\Sigma} = \{\phi \in H^2 | \ \|\phi\|_{\tilde{\Sigma}} < +\infty\}$, where

$$(3.9) \quad \|\phi\|_{\tilde{\Sigma}} = |||S(t)\phi||| = |\phi|_{H^2} + \sup_{t \in \mathbb{R}} \left[(1 + |t|)^{3\left(\frac{1}{2} - \frac{1}{p}\right)} |S(t)\phi|_{W^{1,p}}\right]$$

By (3.3) if $\phi \in H^2 \cap W^{2,p'}$, $\frac{1}{p'} + \frac{1}{p} = 1$, we have $\phi \in \tilde{\Sigma}$ and

$$\|\phi\|_{\tilde{\Sigma}} \leq c\left(|\phi|_{H^2} + |\phi|_{W^{2,p}}\right).$$

The next result has a proof similar to that of theorem 2.1, based on lemma 3.1:

Theorem 3.1: Assume $\psi_0 \in \mathbb{H}^2$ and suppose that, for $p \in]3, \frac{10}{3}]$, $\| \psi_0 \|_{\tilde{\Sigma}}$ is small enough. Then, there exists an unique solution ψ of the equation (1.1) (m>0) verifying $\psi(0) = \psi_0$ and such that $\psi \in C(\mathbb{R}; \mathbb{H}^2) \cap C^1(\mathbb{R}; \mathbb{H}^1)$. Furthermore, $\||\psi\|| < +\infty$.

References

[1] J. Bergh and J. Löfström, Interpolation spaces. An introduction, Springer-Verlag, Berlin, 1976.

[2] Ph. Brenner, On Scattering and Everywhere Defined Scattering Operators for Nonlinear Klein-Gordon Equations, J. Diff. Eq. 56 (1985), 310-344.

[3] J.P. Dias and M. Figueira, Global existence of solutions with small initial data in H^s for the massive nonlinear Dirac equations in three space dimensions, Boll. U.M.I. (7), 1-B(1987), to appear.

[4] J.P. Dias and M. Figueira, Sur l'existence de solution globale pour une équation de Dirac non linéaire avec masse nulle, Comptes Rendus Acad. Sc. Paris, Série I, to appear.

[5] H. Pecher, L^p - Abschätzungen und Klassische Lösungen für nichtlineare Wellengleichungen I, Math. Z. 150 (1976), 159-183.

[6] H. Pecher, Nonlinear Small Data Scattering for the Wave and Klein-Gordon Equation, Math. Z. 185 (1984) 261-270.

[7] M. Reed, Abstract Non-Linear Wave Equations, Lecture Notes in Math. 507, Springer-Verlag, Berlin, 1976.

THE CAUCHY PROBLEM FOR THE
NON-LINEAR KLEIN-GORDON EQUATION

J. Ginibre

Laboratoire de Physique Théorique et Hautes Energies*
Université de Paris XI, Bât. 211, 91405 Orsay, FRANCE

G. Velo

Dipartimento de Fisica
Universita di Bologna and INFN
Sezione di Bologna, ITALY

1. INTRODUCTION

The purpose of this lecture is to summarize some recent perturbative results on the Cauchy problem for the non linear Klein-Gordon equation

$$\Box \varphi \equiv \ddot{\varphi} - \Delta \varphi = -f(\varphi). \tag{1.1}$$

Here φ is a complex valued function defined in space time $\mathbb{R}^{n+1}$, the upper dot denotes the time derivative, Δ is the Laplace operator in $\mathbb{R}^n$ and f is a non linear complex valued function. Only the case $n \geq 2$ will be considered since the special case $n = 1$ is simpler and would require slightly modified statements.

The existence and the properties of the solutions φ depend in a crucial way on the initial data and on the non linear term f. The situation can be best illustrated by considering the following typical form of f

$$f(\varphi) = \lambda_0 \varphi + \lambda \varphi \, |\varphi|^{p-1} \tag{1.2}$$

with λ_0, $\lambda \in \mathbb{R}$ and $1 < p < \infty$. In this case, if one restricts conveniently the space where to operate and therefore the space of initial data, the Cauchy problem for the equation (1.1) has a unique local (in time) solution. The extension of a local solution to a global one requires, in general, additional assumptions since, as it appears already in elementary examples, local solutions may fail to exist beyond a finite time. The standard procedure to prevent this kind of phenomenon makes

---

*Laboratoire associé au Centre National de la Recherche Scientifique.

use, in an essential way, of the conservation of the energy and of some of its positivity properties. For this reason, while there is a large flexibility in the choice of the space where to solve the local Cauchy problem, the available proofs of existence of solutions for the global Cauchy problem always request the initial data to belong to the energy space $X_e \equiv (H^2(\mathbb{R}^n) \cap L^{p+1}(\mathbb{R}^n)) \oplus L^2(\mathbb{R}^n)$. These solutions will be called finite energy solutions. The positivity property of the energy depends crucially on the sign of λ : if $\lambda < 0$, finite energy solutions, known to exist for short time, can blow up; if $\lambda \geq 0$, any element of X_e can be taken as the initial condition of a global solution in X_e.

Existence of global solutions of the Cauchy problem in X_e for $\lambda \geq 0$ can be obtained by a compactness technique, which is inherently non perturbative in character. Instead, all proofs of uniqueness require a perturbative argument which can be implemented in X_e under the assumption

$$p - 1 < 4/(n - 2) . \qquad (1.3)$$

The condition (1.3) can be relaxed at the price of proving uniqueness in a space smaller than X_e, in which case, however, the spaces of existence and uniqueness do not match any more.

Here a presentation is given to the chain of perturbative arguments which leads to the existence and uniqueness in X_e of global solutions of the Cauchy problem for the equation (1.1) under the assumption (1.3) in the situation where the energy is suitably bounded from below ($\lambda \geq 0$ for the example (1.2)) and for any $n \geq 2$. Proofs of the statements, details and references to previous work are given in [3] . Additional useful information can be found in [2] .

We now provide the main notation. We denote by $\| \ \|_r$ the norm in $L^r = L^r(\mathbb{R}^n)$. With each r it is convenient to associate the variables $\gamma(r)$ and $\delta(r)$ defined by

$$\gamma(r)/(n - 1) = \delta(r)/n = 1/2 - 1/r .$$

For each integer k we denote by $H^k \equiv H^k(\mathbb{R}^n)$ the usual Sobolev spaces. We shall use the homogeneous Besov spaces of arbitrary order and the associated Sobolev inequalities, for which we refer to the Appendix both for information and references. We use the notation $\dot{B}_r^\rho \equiv \dot{B}_{r2}^\rho(\mathbb{R}^n)$ for those spaces. For any interval I, for any Banach space B, we denote by $\mathscr{C}(I, B)$ the space of strongly continuous functions from I to B. For any q, $1 \leq q \leq \infty$, we denote by $L^q(I, B)$ (resp. $L_{loc}^q(I, B)$),

the space of measurable functions φ from I to B such that $\| \varphi(\cdot); B \|$
$\in L^q(I)$ (resp. $\| \varphi(\cdot); B \| \in L^q_{loc}$ (I)).

We shall need the operators $K(t) = (-\Delta)^{-1/2} \sin(-\Delta)^{1/2}t$ and
$\dot{K}(t) = \cos(-\Delta)^{1/2}t$; both are bounded and strongly continuous with
respect to t in H^k for any k.

2. THE LOCAL CAUCHY PROBLEM

The study of the Cauchy problem for the equation (1.1) can be
conveniently replaced by the study of the integral equation

$$\varphi = A(t_0, \varphi^{(0)}; \varphi) \tag{2.1}$$

where $t_0 \in \mathbb{R}$, $\varphi^{(0)}$ is a solution of the free wave equation

$$\Box \varphi^{(0)} = 0 \tag{2.2}$$

and

$$A(t_0, \varphi^{(0)}; \varphi) \equiv \varphi^{(0)} + F(t_0; \varphi)$$

with

$$(F(t_0; \varphi))(t) \equiv -\int_{t_0}^{t} d\tau \, K(t - \tau) \, f(\varphi(\tau)) \, . \tag{2.3}$$

At a formal level any solution of the equation (2.1) is a solution of
the equation (1.1) and, conversely, any solution of (1.1) solves (2.1)
with a suitable $\varphi^{(0)}$ (solution of (2.2)) which contains the information
on the initial data at time t_0. The integral (2.3) and the subsequent
ones may be understood in various senses. Generally speaking, when the
functions involved are sufficiently regular, they are ordinary integrals
and the estimates they satisfy determine their extension to a larger
class of functions. Their meaning should be understood from the context
and will not be mentioned explicitly.

In order to prove uniqueness of the solutions of the equation
(2.1), we build a space based on two norms chosen in such a way that
$F(t_0, .)$, restricted to the bounded sets of the first one, is contract-
ing with respect to the second one. Those norms involve space time
integrations and are suggested by the following fundamental estimate
for $K(t)$ [8][6][4]

$$\|K(t)u\|_r \leq C|t|^{1-\delta(r) + \delta(s)} \|u\|_s \qquad (2.4)$$

which holds for

$$0 \leq \delta(r) - \delta(s) \leq \text{Min}\{1 + \gamma(r), \ n(1 - \gamma(r))\}$$
$$1 < s, \ r < \infty \qquad \text{if } n = 2 \ .$$

This motivates the following definitions. For any interval I and suitable values of ℓ, q, r and q_1, we define the spaces

$$\mathcal{X}_0(I) = L^q(I, \ L^\ell) \ , \quad \mathcal{X}_1(I) = L^{q_1}(I, \ L^r)$$

and similarly the local ones in time.

The basic assumption on f is expressed by a power law estimate of the following type

(A1) $f \in \mathcal{C}^1(\mathbb{C}, \mathbb{C})$, $f(0) = 0$ and for some p, $1 \leq p < \infty$ and all $z \in \mathbb{C}$

$$|f'(z)| \equiv \text{Max}\{|\partial f/\partial z| \ , \ |\partial f/\partial \bar{z}|\} \leq C(1 + |z|^{p-1}) \ . \qquad (2.5)$$

The first important property of $F(t_0, .)$ with respect to the previously defined spaces is contained in the next lemma.

Lemma 2.1. Let f satisfy (A1), let I be a bounded interval of time, let $t_0 \in I$ and let $\varphi_1, \varphi_2 \in \mathcal{X}_1(I) \cap \mathcal{X}_0(I)$. Then

$$\|F(t_0, \varphi_1) - F(t_0, \varphi_2); \mathcal{X}_1(I)\| \leq C \|\varphi_1 - \varphi_2; \mathcal{X}_1(I)\|$$
$$\times \{|I|^2 + |I|^{\eta_1} (\sum_{i=1,2} \|\varphi_i; \mathcal{X}_0(I)\|^{p-1})\} \qquad (2.6)$$

provided ℓ, r, q, q_1 satisfy

$$1 \leq \ell, \ r, \ q, \ q_1 \leq \infty$$
$$1 < r < \infty \quad \text{if } n = 2 \ ; \quad |\gamma(r)| \leq 1 \quad \text{if } n \geq 3 \qquad (2.7)$$

$$(p - 1)n/\ell \leq \text{Min} \{1 + \gamma(r), \quad n(1 - \gamma(r))\} \qquad (2.8)$$

$$(p - 1)/q + 1/q_1 \leq 1 \qquad (2.9)$$

$$\eta_1 \equiv 2 - (p - 1)(n/\ell + 1/q) > 0 \ . \qquad (2.10)$$

The proof is a consequence of (2.4) and of Hölder's and Young's inequalities.

Lemma 2.1 implies immediately the following uniqueness result.

<u>Proposition 2.1.</u> Let f satisfy (A1), let ℓ, r, q, q_1 satisfy (2.7)-
(2.10), let I be an open interval, let $t_0 \in I$ and let $\varphi^{(0)} \in \mathcal{X}_{1loc}(I) \cap$
$\mathcal{X}_{0loc}(I)$. Then the equation (2.1) has at most one solution in $\mathcal{X}_{1loc}(I) \cap$
$\mathcal{X}_{0loc}(I)$.

By choosing suitable values of ℓ, r, q, q_1, the conditions (2.7)-
(2.10) can be satisfied for any p; in particular large values of p
require large values of ℓ and q. In this case, in general, finite
energy   solutions do not belong to $\mathcal{X}_{0loc}(\mathbb{R})$ so that the previous
proposition does not apply to this important class of solutions, unless
an upper bound on p is imposed (this bound is expressed by (1.3)).

For the existence of local solutions it is sufficient to show that
balls of arbitrary radius R in $\mathcal{X}_0(I)$ are left invariant by $F(t_0, \cdot)$
for some I depending on R, or, more generally, that this happens for
the balls of an appropriate space $\mathcal{X}_2(I)$ continuously embedded in $\mathcal{X}_0(I)$.
It turns out that a convenient choice for this new space is

$$\mathcal{X}_2(I) = L^q(I, \overset{\circ}{B}{}^\rho_r)$$

for suitable values of ρ, r and q. The Sobolev inequalities (see the
Appendix) imply that $\mathcal{X}_2(I)$ is continuously embedded in $\mathcal{X}_0(I)$ provided
$n/r - \rho = n/\ell$, $\ell \geq 2$ and $\ell \geq r$. In order to show that $F(t_0, \cdot)$ reproduces
the space $\mathcal{X}_2(I)$, the following generalization of (2.4) plays an
important role :

$$\|K(t)u; \overset{\circ}{B}{}^\rho_r\| \leq C|t|^{-\mu}\|u; \overset{\circ}{B}{}^{\rho'}_{r'}\| \tag{2.11}$$

which holds for all r with $0 \leq \lambda(r) \leq 1$ and all ρ, ρ', r', μ such that

$$0 \leq 1 + \mu = \rho + \delta(r) - \rho' - \delta(r') \leq \tfrac{1}{2}(\gamma(r) - \gamma(r'))(1 + 1/\gamma(r)) \leq 1 + \gamma(r).$$

<u>Lemma 2.2.</u> Let f satisfy (A1), let I be a bounded interval of time, let
$t_0 \in I$ and let $\varphi \in \mathcal{X}_2(I)$. Then

$$\|F(t_0,\varphi); \mathcal{X}_2(I)\| \leq C\{|I|^2\|\varphi; \mathcal{X}_2(I)\| + |I|^{\eta_2}\|\varphi; \mathcal{X}_2(I)\|^p\} \tag{2.12}$$

provided ρ, r and q satisfy $0 \leq \rho < 1$ and

$$0 \le \gamma(r) \le (n - 1)/(n + 1) \tag{2.13}$$

$$0 \le (p - 1)(n/r - \rho) \le 1 + \gamma(r) \tag{2.14}$$

$$p \le q \tag{2.15}$$

$$\eta_2 \equiv 2 - (p - 1)(n/r - \rho + 1/q) > 0. \tag{2.16}$$

The proof is similar to that of Lemma 2.1 where (2.4) is replaced by (2.11). The only new ingredient is the Leibnitz rule adapted to fractional derivatives (See Lemma A1).

We can now state the basic local existence and uniqueness result for the solutions of the equation (2.1). We shall denote by $B_2(I, R)$ the closed ball of radius R in $\mathcal{X}_2(I)$.

Proposition 2.2. Let f satisfy (A1), let ρ, r, q and q_1 satisfy $0 \le \rho < 1$, $1 \le q \le q_1 \le \infty$ and (2.13)-(2.16). Then for any R > 0, there exists T(R) > 0 such that, for any $t_0 \in \mathbb{R}$ and for any $\varphi^{(0)} \in B_2(I, R) \cap \mathcal{X}_1(I)$, where $I = [t_0 - T(R), t_0 + T(R)]$, the equation (2.1) has a solution in $B_2(I, 2R) \cap \mathcal{X}_1(I)$ with $\|\varphi; \mathcal{X}_1(I)\| \le 2 \|\varphi^{(0)}; \mathcal{X}_1(I)\|$. That solution is unique in $\mathcal{X}_2(I) \cap \mathcal{X}_1(I)$.

If $n \ge 4$, the assumptions of Proposition 2.2 impose on p the upper limit

$$(p - 1)(n/2 - 3/2 - 1/n) < 2$$

which is obtained by taking $\rho \sim 1$, $\gamma(r) = (n-1)/(n+1)$ and $q = q_1 = \infty$. This condition is not sufficient to ensure that finite energy solutions belong to $\mathcal{X}_2(\mathbb{R})$ since ρ, r and q are too large. A more stringent upper limit on p (given by (1.3)) is necessary in order to allow values of ρ, r, q which permit to accomodate finite energy solutions in $\mathcal{X}_2(\mathbb{R})$.

We now discuss the behaviour of the finite energy solutions. We recall that the energy space X_e has been already defined in the Introduction. If p satisfies (1.3), the Sobolev inequalities imply that

$$X_e = \{(\varphi_0, \psi_0) : \varphi_0 \in H^1, \psi_0 \in L^2\} = H^1 \oplus L^2. \tag{2.17}$$

This is the expression to which from now on we will refer for X_e since in all what follows (1.3) will be always supposed to hold.

With any $(\varphi_0, \psi_0) \in X_e$ we can construct the finite energy solution

$$\varphi^{(0)}(t) = \dot{K}(t - t_0)\varphi_0 + K(t - t_0)\psi_0 \qquad (2.18)$$

of the equation (2.2). This solution belongs to $\mathcal{B}(\mathbb{R}, H^1)$ and its space-time integrability properties are expressed by the following lemma [7][9][5][2] .

<u>Lemma 2.3</u> Let ρ, r, q satisfy

$$0 \leq \delta(r) \leq n/2$$

$$-1 \leq \sigma = \rho + \delta(r) - 1 < 1/2 \qquad (2.19)$$

$$\sigma \leq \gamma(r)/2$$

$$1/q = \text{Max}(0,\sigma) \qquad (2.20)$$

Then, for any $(\varphi_0, \psi_0) \in X_e$, $\varphi^{(0)}$, as defined by (2.18), belongs to $X_2(\mathbb{R})$ and satisfies the estimate

$$\| \varphi^{(0)}; X_2(\mathbb{R})\| \leq C(\|\psi_0\|_2 + \|\nabla\varphi_0\|_2) . \qquad (2.21)$$

This leads to the basic local existence and uniqueness result concerning finite energy solutions.

<u>Proposition 2.3</u> Let f satisfy (A1) and (1.3). Then
(1) There exist ρ, r and q satisfying $0 \leq \rho < 1$, (2.13)-(2.16) and (2.19) and (2.20).
Let X_1 and X_2 correspond to the previous values of ρ, r, q and to $q_1 \geq q$. Then
(2) For any $(\varphi_0, \psi_0) \in X_e$ there exists $T > 0$ depending only on $\|(\varphi_0, \psi_0); X_e\|$ such that for any $t_0 \in \mathbb{R}$, the equation (2.1) with $\varphi^{(0)}$ defined by (2.18) has a unique solution in $X_1(I) \cap X_2(I)$ where $I = [t_0 - T, t_0 + T]$.
(3) For any $(\varphi_0, \psi_0) \in X_e$, for any interval I, for any $t_0 \in I$, the equation (2.1) with $\varphi^{(0)}$ defined by (2.18) has at most one solution in $X_{1loc}(I) \cap X_{2loc}(I)$.

3. The Global Cauchy problem

Once the local Cauchy problem is solved, the next natural question consists in ascertaining whether the solutions of the equation (1.1) obtained by the previous perturbative technique can be extended to all times. This continuation is possible if we can find an a priori bound on the norms of the local solutions and, for this purpose, the energy is a fundamental quantity, provided it satisfies an appropriate positivity condition. The relevant assumption can be formulated in the following way :

(A2) There exists a function $V \in \mathcal{C}^1(\mathbb{C}, \mathbb{R})$ such that $V(0) = 0$, $V(z) = V(|z|)$ for all $z \in \mathbb{C}$ and $f(z) = \partial V / \partial \bar{z}$. For all $R > 0$, V satisfies the estimate

$$V(R) \geq -a^2 R^2 \tag{3.1}$$

for some $a \geq 0$.

For $(\varphi, \psi) \in X_e$ and such that $V(\varphi) \in L^1$ the energy is defined by

$$E(\varphi, \psi) = \|\psi\|_2^2 + \|\nabla\varphi\|_2^2 + \int dx \, V(\varphi(x)) . \tag{3.2}$$

The first part of (A2), namely the relation between f and V, implies, at least at the formal level, that

$$\frac{d}{dt} E(\varphi(t), \dot{\varphi}(t)) = 0 \tag{3.3}$$

where φ is a solution of the equation (1.1), whereas the lower bounded-ness condition (3.1) prevents that an infinite compensation between the kinetic and potential parts of the energy takes place. Those two facts altogether yield a uniform bound for φ in H^1 at finite times in terms of the initial data. The actual proof of energy conservation proceeds through the following path. One first introduces suitable cut-offs both in the equation and in the initial data, then proves a regularized form of the conservation law, and finally deduces (3.3) by removing the cut-offs. The next proposition summarizes the situation in exact terms.

<u>Proposition 3.1.</u> Let f satisfy (A1), (1.3) and (A2). Let $(\varphi_0, \psi_0) \in X_e$ let I be an open interval and let $t_0 \in I$. Let ρ, r and q satisfy $0 \leq \rho < 1$, (2.13)-(2.15), (2.19), (2.20) and let $q_1 \geq q$. Let $\varphi^{(0)}$ be defined by

(2.18) and let φ be a solution of the equation (2.1) in $\mathcal{X}_1(I) \cap \mathcal{X}_2(I)$. Then $(\varphi, \dot{\varphi}) \in \mathcal{C}(I, H^1 \oplus L^2)$ and φ satisfies the conservation of the energy

$$E(\varphi(t), \dot{\varphi}(t)) = E(\varphi_0, \psi_0) \equiv E \qquad (3.4)$$

and the estimates

$$\|\varphi(t)\|_2 \leq e(E, t - t_0) \qquad (3.5)$$

$$\|\dot{\varphi}(t)\|_2^2 + \|\nabla\varphi(t)\|_2^2 \leq \dot{e}(E, t - t_0)^2 \qquad (3.6)$$

where

$$e(E, \tau) = \|\varphi_0\|_2 \cosh a|\tau| + (E + a^2 \|\varphi_0\|_2^2)^{1/2} a^{-1} \sinh a|\tau|$$

We can now state the basic global existence and uniqueness result for finite energy solutions.

<u>Proposition 3.2.</u> Let f satisfy (A1, (1.3) and (A2). let $(\varphi_0, \psi_0) \in X_e$, let $t_0 \in \mathbb{R}$ and let $\varphi^{(0)}$ be defined by (2.18). Then the equation (2.1) has a unique solution φ such that $(\varphi, \dot{\varphi}) \in \mathcal{C}(\mathbb{R}, X_e)$. That solution satisfies the conservation of the energy (3.3) and the bounds (3.4) and (3.5).

Let ρ, r, q and q_1 be as in Proposition 3.1. Then the solution is unique in $\mathcal{X}_{1loc}(\mathbb{R}) \cap \mathcal{X}_{2loc}(\mathbb{R})$.

The proof is an immediate consequence of previous results. The existence of global solutions follows from Proposition 2.3 and the a priori estimates of Proposition 3.1, the uniqueness in $\mathcal{X}_{1loc}(\mathbb{R}) \cap \mathcal{X}_{2loc}(\mathbb{R})$ follows from Proposition 2.2 and the uniqueness in $\mathcal{C}(\mathbb{R}, X_e)$ follows from the fact that any solution in $L^\infty_{loc}(\mathbb{R}, X_e)$ belongs to $\mathcal{X}_{1loc}(\mathbb{R}) \cap \mathcal{X}_{2loc}(\mathbb{R})$ (see Lemma 3.3 of [2]).

APPENDIX

In this appendix we collect some basic facts about the homogeneous Besov and Sobolev spaces. Additional information can be found in [1] , [10] and in the Appendix of [2].

We define the closed subspace $Z = Z(\mathbb{R}^n)$ of $\mathcal{S} = \mathcal{S}(\mathbb{R}^n)$ by

$$Z = \{u \in \mathcal{S} \; : \; (D^\alpha \hat{u})(0) = 0 \text{ for any multiindex } \alpha\} \;.$$

The dual of the inclusion map $Z \subset \mathcal{S}$ is a surjection π from $\mathcal{S}'$ to Z' the kernel of which are the polynomials P, so that $Z' = \mathcal{S}'/P$.

Let now $\psi \in \mathcal{C}_0^\infty(\mathbb{R}^n)$, with $0 \leq \hat{\psi} \leq 1$, $\hat{\psi}(\xi) = 1$ for $|\xi| \leq 1$ and $\hat{\psi}(\xi) = 0$ for $|\xi| \geq 2$, and define for any $j \in \mathbb{Z}$

$$\hat{\varphi}_j(\xi) = \hat{\psi}(2^{-j}\xi) - \hat{\psi}(2^{-(j+1)}\xi) \;.$$

With any $u \in \mathcal{S}'$ we can associate the sequence of functions $\{\varphi_j * u \equiv u_j\}$, $j \in \mathbb{Z}$. That sequence is actually defined for $u \in Z'$ since $u_j = 0$ for $u \in P$. For any $\rho \in \mathbb{R}$ and any r and s with $1 \leq r, s \leq \infty$, we define the homogeneous Besov spaces

$$\dot{B}_{rs}^{\rho} = \{u \in Z' : \{\sum_j 2^{j\rho s} \|\varphi_j * u\|_r^s\}^{1/s} \equiv \|u; \dot{B}_{rs}^{\rho}\| < \infty\}$$

and the useful auxiliary spaces

$$\dot{F}_{rs}^{\rho} = \{u \in Z' : \|\{\sum_j |2^{j\rho}\varphi_j * u|^s\}^{1/s}\|_r \equiv \|u; \dot{F}_{rs}^{\rho}\| < \infty\} \;,$$

with obvious modifications if $s = \infty$. The factor 2^ρ in the sums mimicks a derivative of order ρ. All those spaces are Banach spaces and satisfy the following continuous embeddings

$$\dot{F}_{rs}^{\rho} \subset \dot{B}_{rs}^{\rho}$$

if $1 \leq r \leq s \leq \infty$, and

$$\dot{B}_{rs}^{\rho} \subset \dot{F}_{rs}^{\rho}$$

if $1 \leq s \leq r \leq \infty$. This result allows one to compare the homogeneous Besov spaces to the homogeneous Sobolev spaces to be defined below.

For any $\rho \in \mathbb{R}$, for any $u \in Z$ we define the operator ω^ρ by

$$\widehat{\omega^\rho u}(\xi) = |\xi|^\rho \hat{u}(\xi) \;.$$

One sees by inspection that ω^ρ maps Z into Z and therefore by duality, Z' into Z'. The space $X \equiv \bigcup_{1 \leq r < \infty} L^r$ is embedded in $\mathcal{S}'$ and therefore in Z' so that, since any $u \in \pi(X)$ is the image of only one element in X, the space X is canonically embedded in Z'. For any ρ, for any r with $1 \leq r < \infty$, we define the homogeneous Sobolev space $\dot{H}_r^\rho$ by

$$\dot{H}_r^\rho = \omega^{-\rho} L^r \quad .$$

Equipped with the norm $\|u; \dot{H}_r^\rho\| = \|\omega^\rho u\|_r$, $\dot{H}_r^\rho$ becomes a Banach space. A non trivial argument shows that $\dot{F}_{r2}^\rho = \dot{H}_r^\rho$ for $1 < r < \infty$, so that the previous embeddings imply

$$\dot{B}_{r2}^\rho \subset \dot{H}_r^\rho \subset \dot{B}_{rr}^\rho$$

if $2 \leq r < \infty$, and

$$\dot{B}_{rr}^\rho \subset \dot{H}_r^\rho \subset \dot{B}_{r2}^\rho$$

if $1 < r \leq 2$.

For our purposes the useful Sobolev inequalities in Besov spaces take the form of the following continuous inclusion

$$\dot{B}_{r2}^\rho \subset \dot{B}_{\ell 2}^\sigma$$

with $n/r - \rho = n/\ell - \sigma$, and $1 \leq r \leq \ell \leq \infty$. In particular

$$\dot{B}_{r2}^\rho \subset L^\ell$$

with $n/r - \rho = n/\ell$, $2 \leq \ell$ and $1 \leq r \leq \ell < \infty$.

Finally we quote a result of great use in dealing with non linear problems in Besov spaces (see Lemma 3.2 of [2]) :

<u>Lemma A.1.</u> Let $f \in \mathcal{C}^1(\mathbb{C}, \mathbb{C})$ with $|f'(z)| \leq C |z|^{p-1}$ for some p, $1 \leq p < \infty$. Let $0 < \rho < 1$, let $1 \leq r \leq k \leq \infty$ and $1/s = 1/r - 1/k$. then the following inequality holds

$$\|f(u); \dot{B}_{r2}^\rho\| \leq C \|u; \dot{B}_{k2}^\rho\| \| |u|^{p-1}\|_s$$

for all u such that $u \in L^m$ for some m, $1 \leq m < \infty$ and such that the norms in the right hand side are finite.

REFERENCES

1. Bergh, J., Löfström, J. : Interpolation spaces. Berlin-Heidelberg
    -New York : Springer 1976.
2. Ginibre, J., Velo, G. : The Global Cauchy problem for the non linear
    Klein-Gordon equation. Math. Z. <u>189</u>, 487-505 (1985).
3. Ginibre, J., Velo, G. : The Global Cauchy problem for the non linear
    Klein-Gordon equation-II.To appear in Ann. I.H.P. (Anal. non lin.)
4. Marshall, B., Strauss, W., Wainger, S. : L^p-L^q estimates for the
    Klein-Gordon equation. J. Math. Pur. Appl. <u>59</u>, 417-440 (1980).
5. Pecher, H. : Non linear small data scattering for the wave and
    Klein-Gordon equation. Math. Z. <u>185</u>, 261-270 (1984).
6. Peral, J.C. : L^p-estimates for the wave equation. J. Funct. Anal.
    <u>36</u>, 114-145 (1980).
7. Segal, I.E. : Space-time decay for solution of wave equations.
    Adv. Math. <u>22</u>, 305-311 (1976).
8. Strichartz, R. : Convolutions with kernels having singularities on
    a sphere. Trans. Amer. Math. Soc. <u>148</u>, 461-470 (1970).
9. Strichartz, R. : Restrictions of Fourier transforms to quadratic
    surfaces and decay of solutions of wave equations. Duke Math.I.
    <u>44</u>, 705-714 (1977).
10. Triebel H. : Theory of function spaces. Basel : Birkhaüser 1983.

CONFORMAL INVARIANCE AND TIME DECAY
FOR NONLINEAR WAVE EQUATIONS

J. Ginibre

Laboratoire de Physique Théorique et Hautes Energies[*]

Université de Paris XI, bâtiment 211, 91405 Orsay, FRANCE

G. Velo

Dipartimento di Fisica

Università di Bologna and INFN, Sezione di Bologna, ITALY

1. INTRODUCTION AND GENERAL BACKGROUND

In this lecture, we describe some implications of the approximate conformal invariance of the nonlinear wave (NLW) equation on the time decay of its solutions. The equation under study is

$$\Box \varphi \equiv \ddot{\varphi} - \Delta \varphi = -f(\varphi) \tag{1.1}$$

where φ is a complex valued function defined in space-time $\mathbb{R}^{n+1}$, the upper dot denotes the time derivative, Δ is the Laplace operator in $\mathbb{R}^n$, and f is a nonlinear complex valued function, a typical form of which is

$$f(\varphi) = \lambda \varphi |\varphi|^{p-1} \tag{1.2}$$

with $1 \leq p < \infty$. The question under study is a special case of the more general question of determining the asymptotic behaviour in time of the solutions of an evolution equation for which the global Cauchy problem can be solved in a reasonably general fashion. It is known that an important tool for that purpose consists of a priori estimates derived from exact or approximate conservation laws associated with the equation. Another example which will be considered here for comparison, is that of the nonlinear Schrödinger (NLS) equation

$$i \dot{\varphi} = -\frac{1}{2} \Delta \varphi + f(\varphi) \tag{1.3}$$

---

[*]Laboratoire associé au Centre National de la Recherche Scientifique.

with a similar interaction f, together with its approximate pseudo-conformal conservation law [4].

It is known (see [10] and references therein contained) that for initial data ($\varphi(0) = \varphi_0$, $\dot{\varphi}(0) = \psi_0$) in the energy space $X_e \equiv H^1 \oplus L^2$, the global Cauchy problem for the equation (1.1) has a unique solution φ which is a continuous function of time with values in X_e under assumptions on f that reduce to $\lambda \geq 0$ and $p < 1 + 4/(n-2)$ in the special case (1.2). Furthermore φ satisfies the conservation of the energy

$$E(\varphi(t), \quad \dot{\varphi}(t)) = E(\varphi_0, \psi_0) \tag{1.4}$$

for all $t \in \mathbb{R}$, where

$$E(\varphi, \Psi) = \| \psi \|_2^2 + \| \nabla \varphi \|_2^2 + \int dx\ V(\varphi(x)) \quad, \tag{1.5}$$

$\| \cdot \|_2$ denotes the norm in $L^2 \equiv L^2(\mathbb{R}^n)$ and $V(\varphi) = 2\lambda(p+1)^{-1}|\varphi|^{p+1}$ in the special case (1.2)(See the general definition in Section 2 below). Similar results hold for the NLS equation (1.3).

Conservation laws for equations such as (1.1) and (1.3) formally result from the fact that the equation (i) is derived from a variational problem and (ii) is invariant under a one parameter continuous group of transformations, through the use of Noether's theorem (see [3] and references therein quoted). One is generally given a Lagrangian density $\mathcal{L} (\varphi, \{\partial_\mu \varphi\})$ where $\{\partial_\mu \varphi; 0 \leq \mu \leq n\} \equiv \{\partial_0 \varphi \equiv \dot{\varphi}, \{\partial_j \varphi; 1 \leq j \leq n\} \equiv \nabla \varphi\}$ is the set of first order derivatives of φ, and the equation is the Euler-Lagrange equation associated with the action A_Λ defined for any open subset $\Lambda \subset \mathbb{R}^{n+1}$ with smooth boundary by

$$A_\Lambda = \int \mathcal{L} (\varphi, \{\partial_\mu \varphi\}) dt\ dx .$$

With a one parameter transformation group of $\mathbb{R}^{n+1}$ under which the equation is invariant, there is associated an infinitesimal transformation of $\mathbb{R}^{n+1}$, namely a vector field in $\mathbb{R}^{n+1}$ $\{\delta x^\mu; 0 \leq \mu \leq n\} = \{\delta x^0 \equiv \delta t, \{\delta x^j; 1 \leq j \leq n\} \equiv \delta x\}$ and an infinitesimal transformation of the field $\delta \varphi$. If φ satisfies the variational equation, the infinitesimal transformation of the action is easily seen to be

$$\delta A_\Lambda = - \int_{\partial \Lambda} J^\mu\ d\sigma_\mu$$

where $d\sigma_\mu$ is the area n-form in $\mathbb{R}^{n+1}$, J^μ is the current vector

$$J^\mu = -\mathscr{L}\,\delta x^\mu - [\partial\mathscr{L}/\partial(\partial_\mu\varphi)]\delta\varphi \quad,$$

we use the summation convention of Relativity Theory, and the Minkowski metric $g_{\lambda\mu}$ used to raise and lower indices is defined by $g_{00} = -g_{ii} = 1$, $g_{\lambda\mu} = 0$ for $\lambda \neq \mu$. If the action is invariant, namely $\delta A_\Lambda = 0$ for any Λ, then the current J^μ is conserved, namely $\partial_\mu J^\mu = 0$. Taking for Λ the region $s \leq x^0 \leq t$ and assuming sufficient decay at infinity in space, one obtains the conservation law $Q(t) = Q(s)$ for all s and t, where the charge $Q(t)$ is defined by

$$Q(t) = \int_{x^0=t} J^0\,dx \;.$$

If the invariance is only approximate, one obtains only the approximate conservation law

$$Q(t) - Q(s) = \int_s^t R(\tau)\,d\tau \tag{1.6}$$

with

$$R(\tau) = \int_{x^0=\tau} \partial_\mu J^\mu\,dx \;.$$

In the case of the NLW equation (1.1), the Lagrangian density is $\mathscr{L}(\varphi,\{\partial_\mu\varphi\}) = |\dot\varphi|^2 - |\nabla\varphi|^2 - V(\varphi) \equiv \partial^\mu\overline\varphi\,\partial_\mu\varphi - V(\varphi)$, where V is the same potential function as in the energy (1.5). The relevant transformations of space-time are those of the conformal group, generated by the Poincaré transformations, space-time dilations, and the transformations obtained therefrom by conjugation under the external automorphism $x^\mu \to x^\mu(x_\lambda x^\lambda)^{-1}$. Of special interest are the transformations thereby obtained from space-time translations. The corresponding infinitesimal transformation of space-time is given by

$$\delta x^\mu = -2a^\lambda x_\lambda x^\mu + x^\lambda x_\lambda a^\mu$$

where $a = \{a^\mu\}$ is a space-time vector, and the transformation of the field φ is given by

$$\delta\varphi = -\delta x^\lambda \partial_\lambda\varphi + (n-1)a^\lambda x_\lambda\varphi \;.$$

Taking in particular for a the vector $\{1, 0, \ldots, 0\}$, one obtains the relevant conservation law in the form (1.6) with

$$Q(t, \varphi, \dot{\varphi}) = Q_0(t, \varphi, \dot{\varphi}) + Q_1(t, \varphi) \quad , \tag{1.7}$$

$$Q_0(t, \varphi, \psi) = \|r\psi\|_2^2 + t^2 \|\psi\|_2^2 + \|r\nabla\varphi\|_2^2 + t^2 \|\nabla\varphi\|_2^2$$
$$+ 2t \operatorname{Re} \langle\psi, 2x.\nabla\varphi + (n-1)\varphi\rangle - (n-1)\|\varphi\|_2^2 \quad , \tag{1.8}$$

$$Q_1(t, \varphi) = \int dx (t^2 + r^2) V(\varphi) \quad , \tag{1.9}$$

$$R(t, \varphi) = 2t \int dx\, W(\varphi) \quad , \tag{1.10}$$

$$W(\varphi) = (n+1) V(\varphi) - (n-1) \operatorname{Re} \overline{\varphi} f(\varphi) \quad , \tag{1.11}$$

where $r = |x|$; $\langle.,.\rangle$ denotes the scalar product in L^2. Q as given by (1.7)-(1.9) will be called the conformal charge, and Q_0 and Q_1 will be called the kinetic and potential parts thereof. By an elementary computation, Q_0 can be rewritten in the form

$$Q_0(t, \varphi, \psi) = \|t\nabla\varphi + x\psi\|_2^2 + \|L\varphi\|_2^2 + \|x.\nabla\varphi + t\psi + (n-1)\varphi\|_2^2 \tag{1.12}$$

where $L = x \times \nabla$ is the angular momentum operator, which shows in particular that Q_0 is non negative.

The fact that the conformal conservation law implies some time decay of the solutions appears immediately in the case of the single power interaction (1.2), where $V(\varphi) = 2\lambda (p+1)^{-1} |\varphi|^{p+1}$ and

$$W(\varphi) = [n+1 - (p+1)(n-1)/2] V(\varphi) \quad . \tag{1.13}$$

In fact for $\dot{\lambda} \geq 0$ and $p+1 \geq \ell_1 \equiv 2(n+1)/(n-1)$, or equivalently $p \geq 1 + 4/(n-1)$, one has $W(\varphi) \leq 0$, $t R(t, \varphi) \leq 0$ and therefore $Q(t) \leq Q(0)$ for all $t \in \mathbb{R}$. Since $Q_0 \geq 0$, one obtains by using only Q_1

$$\|\varphi(t)\|_{p+1} \leq c\, t^{-2/(p+1)} \tag{1.14}$$

where $\|.\|_\ell$ denotes the norm in $L^\ell \equiv L^\ell(\mathbb{R}^n)$. That decay property was used extensively in previous works on the NLW equation [24].

In order to assess the strength of the decay property (1.14) it is useful to compare it with the decay available for solutions of the free wave equation $\Box \varphi = 0$, for which one obtains at best

$$\|\varphi(t)\|_\ell \leq C(1+|t|)^{-\gamma(\ell)} \tag{1.15}$$

with

$$\gamma(\ell) = (n-1)(1/2-1/\ell) \ . \tag{1.16}$$

That the decay (1.15) is optimal is easily seen in the special case of the dimension n = 3 where the solution of $\Box \varphi = 0$ with initial data ($\varphi_0 = 0$, ψ_0) is given by

$$\varphi(t) = (4\pi t)^{-1} \delta(r-t) * \psi_0$$

where $*$ denotes the convolution in $\mathbb{R}^n$, and can be explicitly computed for $\psi_0 = \chi(r \leq a)$, the characteristic function of the ball of radius a, to be

$$\varphi(t) = (4r)^{-1} [a^2-(t-r)^2]_+$$

for $t \geq a$. More generally solutions of $\Box \varphi = 0$ with the decay (1.15) can be constructed with the help of the following lemma. There we use the notation $\omega = \sqrt{-\Delta}$, and in addition to $\gamma(\ell)$ defined by (1.16), we also use $\alpha(\ell)$ and $\delta(\ell)$ defined by

$$\alpha(\ell) \equiv n^{-1}\delta(\ell) \equiv 1/2 - 1/\ell \ . \tag{1.17}$$

<u>Lemma 1.1</u> [17, 20, 26]. Let $n \geq 2$, let ℓ and s satisfy

$$1 + \delta(s) \leq \delta(\ell) \leq \text{Min}(1+ \alpha(s), n(1+ \delta(s))) \tag{1.18}$$

and $2 < \ell < \infty$ if n = 2. Then the following estimate holds

$$\| \omega^{-1}e^{\pm i\omega t} \varphi \|_\ell \leq c|t|^{1-\delta(\ell)+\delta(s)} \| \varphi \|_s \tag{1.19}$$

for all $t \neq 0$ and all $\varphi \in L^s$.

In particular, the optimal decay (1.15) for solutions

$$\varphi(t) = K(t)\psi_0 + \dot{K}(t) \varphi_0 \tag{1.20}$$

of $\Box \varphi = 0$, where $K(t) = \omega^{-1} \sin\omega t$ and $\dot{K}(t) = \cos\omega t$, can be obtained for suitable initial data (φ_0, ψ_0) by using Lemma 1.1 in the border-line case $\alpha(\ell) = 1+\delta(s)$.

Since $\gamma(p+1) = 2/(p+1)$ for $p+1 = \ell_1$, the decay (1.14) obtained from the conformal conservation law coincides with the optimal decay (1.15) in that limiting case, while it is generally weaker in the allowed range $p+1 \geq \ell_1$. That result can then be improved (actually has been improved in previous works [24, 29]) by substituting the decay (1.14) in the integral equation associated with the equation (1.1) and performing various estimates thereof. It is however a natural question to ask whether stronger results can be obtained more directly by a more efficient application of the conformal conservation law, using in particular the kinetic part Q_0 of the conformal charge. For that purpose, it is useful to compare the case of the NLW equation with the simpler case of the NLS equation (1.3). The solutions of that equation satisfy the pseudo-conformal conservation law which is similar to the conformal conservation law of the NLW equation [4]. That law takes again the form (1.6) where however (the subscript S stands for Schrödinger) $Q_S = Q_{0S} + Q_{1S}$ with

$$Q_{0S}(t, \varphi) = \frac{1}{2} \, \| (x+it\nabla) \, \varphi \, \|_2^2 \tag{1.8}_S$$

$$Q_{1S}(t, \varphi) = t^2 \int dx \, V(\varphi) \tag{1.9}_S$$

$$R_S(t, \varphi) = t \int dx \, W_S(\varphi) \tag{1.10}_S$$

$$W_S(\varphi) = (n+2) \, V(\varphi) - n \, \text{Re} \, \overline{\varphi} \, f(\varphi). \tag{1.11}_S$$

In the case of a single power interaction (1.2), one obtains

$$W_S(\varphi) = [n+2 - (p+1) \, n/2]V(\varphi) \tag{1.13}_S$$

so that for $\lambda \geq 0$ and $p+1 \geq 2(n+2)/n$ or equivalently $p \geq 1+4/n$, one has as before $W_S(\varphi) \leq 0$, $t \, R_S(t, \varphi) \leq 0$, $Q_S(t) \leq Q_S(0)$ for all t in $\mathbb{R}$, and the solution φ again satisfies the decay (1.14). The optimal time decay available for solutions of the free Schrödinger equation $i \, \dot{\varphi} = (-1/2)\Delta\varphi$ is known to be

$$\| \varphi(t) \|_\ell \leq C(1+|t|)^{-\delta(\ell)} \tag{1.15}_S$$

where $\delta(\ell)$ is defined by (1.17). The decay (1.14) is again identical with the optimal one in the limiting case $p+1 = 2(n+2)/n$, and weaker in the allowed range $p+1 \geq 2(n+2)/n$. In the NLS case however, one can obtain easily a better result by using directly the kinetic part Q_{0S}

of the pseudo-conformal charge. For that purpose we introduce the free Schrödinger group $U(t) = \exp(i(t/2)\Delta)$. For each $t \neq 0$, $U(t)$ can be represented as the operator of   convolution with the function $(2\pi it)^{-n/2}$ $\exp[ix^2/(2t)]$. Furthermore, the following identity holds

$$J \equiv x + it\nabla = U(t) x \, U(-t) \, . \tag{1.21}$$

One can then prove the following estimate.

__Lemma 1.2__ [4]. Let $2 \le \ell \le 2^* \equiv 2n/(n-2)$, $\ell < \infty$, if $n = 2$. Then

$$\| \varphi \|_\ell \le C |t|^{-\delta(\ell)} \| \, x \, U(-t) \, \varphi \|_2^{\delta(\ell)} \| \varphi \|_2^{1-\delta(\ell)} \tag{1.22}$$

for all $t \neq 0$ and all φ for which the right hand side is finite. (Here C is a Sobolev constant).

__Proof__ By an homogeneity argument, it is sufficient to prove the result for $t = -1$. In that case $U(1) = \theta S \, \mathcal{F} \, S$ where $\mathcal{F}$ is the Fourier transform, θ is a constant phase factor, and S is the operator of multiplication by $\exp(ix^2/2)$. Then

$$\| x \, U(1) \, \varphi \|_2 = \| x \, \mathcal{F} S \, \varphi \|_2 = \| \nabla S \, \varphi \|_2$$

by the Plancherel theorem, while

$$\| \varphi \|_\ell = \| S \, \varphi \|_\ell \le C \quad \| \nabla S \, \varphi \|_2^{\delta(\ell)} \quad \| S \, \varphi \|_2^{1-\delta(\ell)}$$

by a standard Sobolev inequality. This proves the lemma.

Q.E.D.

Applying Lemma 1.2 with $\varphi = \varphi(t)$ to a solution of the NLS equation (1.3) yields the optimal decay $(1.15)_S$ for $2 \le \ell \le 2^*$ whenever Q_{0S} is uniformly bounded in time, and in particular in the case of a single power interaction (1.2) with $\lambda \ge 0$ and $p+1 \ge 2(n+2)/n$ considered previously.

The first main result reported in this lecture is the analogue of Lemma 1.2 in the case of the NLW equation, yielding in particular an estimate of the L^ℓ norm of its solutions for $2 \le \ell \le 2^*$ with the optimal time decay (1.15) in terms of the kinetic part Q_0 of the conformal charge (See Proposition 3.2 below and [8] for more details). Actually we shall present a stronger result, since the basic estimate involves space-time weighted L^ℓ- norms and thereby yields an additional decay in space. As in the NLS case, that estimate will emerge as a variant of a Sobolev inequality, but of a much more complicated type than the elementary estimate (1.22).

The previous considerations and results cover the "easy" case where the conformal conservation law immediately implies the boundedness of the conformal charge, corresponding to $p+1 \geq \ell_1$ in the special case (1.2). In order to guess what time decay should be expected in the more difficult case $p+1 < \ell_1$, it is useful to consider again the situation for the NLS equation. The problem of the time decay of solutions can be regarded as part of the general theory of Scattering for that equation [4, 7, 27, 28]. Another part of that theory is the construction of dispersive solutions, namely of solutions that behave asymptotically in time as solutions of the corresponding free equation, obtained by dropping the interaction term f. Dispersive solutions can be constructed by solving the Cauchy problem with infinite initial time by a contraction method, in suitable functional spaces that include some time decay in their definition. The method can be implemented under assumptions on f which reduce to a lower bound on p in the special case (1.2). That lower bound depends on the choice of the time decay that appears in the relevant functional space. If that decay is chosen to be as strong as the optimal decay (1.15)$_S$, the lower bound $p_S(n)$ on p in the NLS case turns out to be the (positive) root of the equation

$$p \, \delta(p+1) = 1 \tag{1.23}$$

or equivalently

$$n \, p^2 - (n+2) \, p - 2 = 0 \, , \tag{1.24}$$

namely $p_S(2) = 1 + \sqrt{2}$, $p_S(3) = 2$, etc. Remarkably enough, it can be shown that the kinetic part of the conformal charge still remains bounded in time and therefore that the optimal decay still holds for all $p > p_S(n)$ [14, 27, 28]. We state the result in the special case (1.2) only, in a somewhat loose way.

<u>Proposition 1.1</u> [14, 27, 28] . Let $p_S(n) < p \leq 2(n+2)/n$, let $\varphi_0 \in H^1$, with $x \, \varphi_0 \in L^2$, and let φ be the (unique H^1-valued) solution of the NLS equation (1.3) with f given by (1.2) and $\lambda \geq 0$, with initial data φ_0 at time zero. Then
(1) $Q_{0S}(t, \varphi(t))$ is uniformly bounded in time.
(2) φ satisfies the decay estimate (1.15)$_S$ for $2 \leq \ell \leq 2^*$ ($\ell < \infty$ if n=2).

Sketch of proof. It suffices to prove part (1), from which part (2) follows by Lemma 1.2. The proof proceeds in two steps.

(1) By reestimating W_S in terms of V in the pseudo-conformal conservation law, one obtains a differential inequality for the pseudo-conformal charge which by integration yields (among other) the decay $(1.15)_S$ for the special value $\ell = p+1$.

(2) One has to estimate $J\varphi$ (see (1.21)) in L^2 uniformly in time. Now J is easily seen to commute with $i\, d/dt + (1/2)\Delta$, so that $J\varphi$ satisfies the differential equation

$$i \frac{d}{dt} J\varphi = -\frac{1}{2} \Delta\, J\varphi + Jf(\varphi) \tag{1.25}$$

and therefore the associated integral equation

$$J\varphi(t) = U(t)x\,\varphi_0 - i \int_0^t d\tau\, U(t-\tau)\, Jf(\varphi(\tau)) . \tag{1.26}$$

Let now $S \equiv S(t) \equiv \exp(ix^2/2t)$. Then J also satisfies the relation

$$J = S\, it\, \nabla S^{-1} \tag{1.27}$$

so that

$$Jf(\varphi) = S\, it\, \nabla\, \overline{S}^{-1} f(\varphi) = S\, it\, \nabla\, f(S^{-1}\varphi)$$

$$= S\, it(\nabla S^{-1}\varphi)\, \frac{\partial f}{\partial\varphi}(S^{-1}\varphi) + S\, it(\overline{\nabla S^{-1}\varphi})\, \frac{\partial f}{\partial\overline{\varphi}}(S^{-1}\varphi)$$

$$= (J\varphi)\, \frac{\partial f}{\partial\varphi} - (\overline{J\varphi})\, \frac{\partial f}{\partial\overline{\varphi}} \quad .$$

Therefore (1.26) is a linear equation for $J\varphi$ regarded as an independent function. One can then use that equation to estimate $J\varphi$ in L^2 uniformly in time by using known estimates on the operator

$$h \to \int_0^t d\tau\, U(t-\tau)h(\tau)$$

[16] together with the estimate $(1.15)_S$ with $\ell = p+1$ obtained in step 1 in order to control the factors $\partial f/\partial\varphi$ and $\partial f/\partial\overline{\varphi}$ coming from $Jf(\varphi)$. That step turns out ot work precisely under the condition $p > p_S(n)$.

Q.E.D.

The preceding line of argument can be extended to some extent to
the NLW equation in the difficult case $p+1 \leq \ell_1$, but one encounters
several difficulties which somehow restrict the scope of the results,
as we now explain. The optimal decay rate (1.15) in the NLW case has
$\delta(\ell)$ replaced by $\gamma(\ell)$ as compared with the NLS case (or for that matter,
with the case of the massive nonlinear Klein-Gordon (NLKG) equation
$\Box \varphi + \varphi + f(\varphi) = 0$). The shift from n to n-1 is best understood by
noticing that dispersion in the massless case $\Box \varphi = 0$ occurs (at least
in odd dimensions by the Huygens principle) in (n-1)-dimensional sub-
manifolds instead of n dimensional space in the massive case and in
the Schrödinger case. Correspondingly, one expects the contraction
argument leading to the proof of existence of dispersive solutions of
the NLW equation (1.1) with interaction (1.2) to work for $p > p_0(n) \equiv$
$p_S(n-1)$ (i.e $p_0(3) = 1 + \sqrt{2}$, $p_0(4) = 2,...$). That lower bound is known
to be optimal, in view of the existing blow up results for attractive
interactions and small initial data in the opposite case $p \leq p_0(n)$
[11, 15, 22]. A natural tool to be used in order to implement the contraction argument is the estimate of Lemma 1.1., which actually produces
the optimal decay (1.15) in the borderline case $\alpha(\ell) = 1 + \delta(s)$. Unfortunately, the use of that lemma produces the expected result only
under the stronger condition $p > p_1(n)$ where $p_1(n) > p_0(n)$ is the
(larger) root of the equation

$$n(n-1)p^2 - p(n^2 + 3n - 2) + 2 = 0 \tag{1.28}$$

[6, 18, 19, 23] . Extending the contraction argument from $p_1(n)$ down
to $p_0(n)$ is a highly non trivial matter and has been done so far only
in dimensions 2 [12] and 3 [15, 21] for special classes of solutions,
by using in an essential way the positivity properties of the propaga-
tor for the free equation $\Box \varphi = 0$, and by using space-time weighted
norms instead of simply L^ℓ-norms.

The general scheme leading to the analogue of Proposition 1.1 can
be implemented in the NLW case; in particular the expression (1.12)
for Q_0 is an adequate substitute for the expression $Q_{0S} = (1/2) \|\vec{J} \varphi\|_2^2$.
Unfortunately the NLW analogue of the first step in the proof of that
proposition does not yield (for $\ell = p+1$) the optimal decay (1.15) but
only the weaker decay

$$\| \varphi(t) \|_\ell \leq C \, |t|^{1-2\delta(\ell)} \quad . \tag{1.29}$$

A simple power counting argument then shows that the remaining part of the proof can be expected to hold only under an assumption on p which allows for the contraction argument mentioned previously to work when only the decay (1.29) is used in the definition of the relevant functional space. That assumption turns out to be $p > p_2(n)$ where $p_2(n) > p_1(n)$ is the (positive) root of the equation [25]

$$p(2\delta(p+1) - 1) = 1 \tag{1.30}$$

or equivalently

$$(n-1)p^2 - (n+2)p - 1 = 0 \ .$$

Thus, although one may conjecture that the analogue of Proposition 1.1 holds in the NLW case for $p > p_0(n)$, the available method of proof is restricted to $p > p_2(n)$.

A final difference between the NLS and NLW cases is the technical fact that the estimates on the integral equation are substantially more difficult in the latter case than in the former. As a consequence the expected results have been proved only for certain dimensions, namely n = 3 and n = 4 [9].The case of higher dimension is technically more complicated and has not been worked out in detail. The case of dimension n = 2 can be analyzed rather completely down to $p_2(2)$, but is plagued with special difficulties in the form of logarithmic factors [8]. The case n = 1 does not seem to be interesting in the present context.

The second main result reported in this lecture is the analogue of Proposition 1.1 in the case of the NLW equation, in the case of space dimension n = 2, 3 and 4 and under assumptions on f that reduce to $p_2(n) < p < 1+4/(n-2)$ in the special case (1.2), according to the preceding discussion.

The remaining part of this lecture is organized as follows. In Section 2 we state the conformal conservation law for the NLW equation with the relevant functional analytic details, including in particular the choice of the relevant space of initial data and the natural assumptions on the interaction f. In section 3, we state the basic apriori estimates in terms of the kinetic part of the conformal charge, extending Lemma 1.2 to the NLW case, and we give a brief sketch of their proof. Finally in Section 4, we derive therefrom the resulting decay estimates of the solutions of the NLW equation, first for general $n \geq 2$ in the easy case corresponding to $p+1 \geq \ell_1$, and then for n = 2, 3, 4 in the harder case $p_2(n) < p \leq \ell_1 - 1$. We refer the reader to [8]

for a complete treatment of the special case n = 2 and to [8,9] for
the details of the proofs.

We conclude this section with the remark that, in the framework of
Scattering Theory, the results of Sections 3 and 4 are the essential
ingredients of the proof of asymptotic completeness (for arbitrarily
large initial data).

2. THE CONFORMAL CONSERVATION LAW.

In order to convert the formal derivation of the conformal conservation law sketched in Section 1 into an actual proof, it is convenient
to apply the following general method. One first regularizes the equation by introducing suitable cut-offs. The solutions of the regularized
equation have sufficient smoothness and decay at infinity in space to
allow for the proof of a regularized version of the conservation law
by the same computations as used in the preceding heuristic derivation.
One finally removes the cut-offs by a limiting procedure.

The regularization uses a local cut-off h and a space cut-off g at
large distances. The cut-off h is taken as a non negative even function
in $\mathcal{C}_o^\infty(\mathbb{R}^n)$, such that $\|h\|_1 = 1$, and the cut-off g as a function in
$\mathcal{C}^1(\mathbb{R}^n)$ with compact support and such that $0 \leq g \leq 1$ and g = 1 in some
region around the origin. Because of the finite propagation speed for
the equation (1.1), the space cut-off can be introduced either in the
initial data or in the interaction. We choose the second possibility
because of the intrinsic interest of localized interactions. We replace
the equation (1.1) by the regularized equation

$$\Box \varphi + h \star g\, f(h \star \varphi) = 0 \qquad\qquad (2.1)$$

and consider the Cauchy problem for (2.1) with regularized initial data
$(\varphi(0), \dot{\varphi}(0)) = (h \star \varphi_0, h \star \psi_0)$.

The limiting procedure consists in letting h tend to a δ function
and g tend to 1 in the following sense : we choose fixed h_1 and g_1 as
described above; for any positive integer j, we define $h_j(x) = j^n h_1(jx)$
and $g_j(x) = g(x/j)$, we take $h = h_j$, $g = g_j$ and we let j tend to infinity.
The solutions of the regularized equation (2.1) tend to solutions of
(1.1) in a sense which makes it possible to take the limit in the conservation law, provided the interaction f and the initial data (φ_0, ψ_0)
satisfy suitable assumptions. These assumptions basically ensure that
(i) the Cauchy problem for (1.1) can be solved in a unique way, and
(ii) the conservation law has a meaning.

The natural condition on (φ, ψ) for the energy $E(\varphi, \psi)$ and the kinetic part of the conformal charge $Q_0(t, \varphi, \psi)$ to be defined is that $(\varphi, \psi) \in \Sigma \equiv \Sigma_1 \oplus \Sigma_0$, where

$$\Sigma_1 = \{\varphi \in L^2 : \nabla \varphi, \ r \nabla \varphi \in L^2\} \tag{2.2}$$

and

$$\Sigma_0 = \{\psi \in L^2 : r \ \psi \in L^2\} \quad . \tag{2.3}$$

The spaces Σ_1 and Σ_0 are Hilbert spaces with norms

$$\| \varphi; \ \Sigma_1 \|^2 = \| \varphi \|_2^2 + \| \nabla \varphi \|_2^2 + \| r \ \nabla \varphi \|_2^2$$

$$\| \psi; \ \Sigma_0 \|^2 = \| \psi \|_2^2 + \| r \psi \|_2^2 \quad .$$

We shall consider only initial data $(\varphi_0, \psi_0) \in \Sigma$.

The assumptions on the interaction f consist of those required to solve the global Cauchy problem with uniqueness (namely (A1) and (A2, a, b) below) supplemented with an assumption which prevents $Q_1(t, \varphi)$ to tend to $-\infty$ while $Q_0(t, \varphi, \psi)$ tends to $+\infty$ for fixed $Q(t, \varphi, \psi)$ (namely (A2c)). They can be stated as follows.

(A1). $f \in \mathscr{C}^1(\mathbb{C}, \mathbb{C})$ and $f(0) = 0$. If $n \geq 2$, f satisfies the estimate

$$|f'(z)| \equiv \operatorname{Max}(|\partial f/\partial z|, |\partial f/\partial \bar{z}|) \leq C(|z|^{p_1-1} + |z|^{p_2-1}) \tag{2.4}$$

for some p_1, p_2 with $1 \leq p_1 \leq p_2 < 1+4/(n-2)$, and all $z \in \mathbb{C}$.

(A2). (a) There exists a function $V \in \mathscr{C}^1(\mathbb{C}, \mathbb{R})$ such that $V(0) = 0$, $V(z) = V(|z|)$ for all $z \in \mathbb{C}$, and $f(z) = \partial V/\partial \bar{z}$.
   (b) V satisfies the estimate

$$V(R) \geq - C \ R^2 \tag{2.5}$$

   for some $C \geq 0$ and all $R \geq 0$.
   (c) V satisfies the estimate

$$V(R) \geq - C \ R^{2+4/n} \tag{2.6}$$

for some $C \geq 0$ and all $R \geq 0$.

We can now state the conservation law, first with the cut-off g still included, and finally with both cut-offs removed.

Proposition 2.1 Let f satisfy (A1) and (A2 a, b), let $(\varphi_0, \psi_0) \epsilon \Sigma$, let $t_0 \epsilon \mathbb{R}$ and let φ be the (unique H^1-valued) solution of (1.1) with initial data (φ_0, ψ_0) at time t_0. Then $(\varphi, \dot{\varphi}) \epsilon \mathscr{C}(\mathbb{R}, \Sigma)$ and for all s, t $\epsilon \mathbb{R}$, the following identity holds:

$$Q_0(t, \varphi(t), \dot{\varphi}(t)) + \int dx (t^2 + r^2) g V(\varphi(t)) =$$

$$Q_0(s, \varphi(s), \dot{\varphi}(s)) + \int dx (s^2 + r^2) g V(\varphi(s))$$

$$+ \int_s^t 2\tau d\tau \int dx \{g W(\varphi(\tau)) + (x.\nabla g) V(\varphi(\tau))\} \quad . \tag{2.7}$$

Proposition 2.2 Let f satisfy (A1) and (A2 a, b, c), let $(\varphi_0, \psi_0) \epsilon \Sigma$. Assume that

$$\int dx \, r^2 \, |V(\varphi_0)| < \infty$$

Let $t_0 \epsilon \mathbb{R}$ and let φ be the (unique H^1-valued) solution of (1.1) with initial data (φ_0, ψ_0) at time t_0. Then $(\varphi, \dot{\varphi}) \epsilon \mathscr{C}(\mathbb{R}, \Sigma)$, the integral $\int dx \, r^2 \, V(\varphi)$ is absolutely convergent for each t $\epsilon \mathbb{R}$ and continuous with respect to t, and φ satisfies the conformal conservation law (1.6) - (1.11) for all s, t $\epsilon \mathbb{R}$, i.e. (2.7) with g = 1.

We refer to [8], Section 2 for the details of the proofs.

3. CONFORMAL ESTIMATES

In this section, we derive some a priori estimates involving the kinetic part of the conformal charge, thereby providing an analogue of Lemma 1.2 in the case of the wave equation. The starting point is the expression (1.12) for $Q_0(t, \varphi, \psi)$, where we regard t as a real positive parameter and φ, ψ as two independent functions in Σ_1 and Σ_0 respective-ly. We use two different methods.

The first method exploits Lemma 1.1. For that purpose, we first rewrite Q_0 in the form

$$Q_0(t, \varphi, \psi) = \| t\nabla\varphi + x\psi \|_2^2 + \| x\omega\varphi - \omega^{-1}t\nabla\psi \|_2^2 \qquad (3.1)$$

with $\omega = \sqrt{-\Delta}$, after an elementary computation involving the commutator of x with ω. We then recombine φ and ψ into two new functions $\varphi_\pm$ which would be the positive and negative frequency parts of φ if φ were a solution of $\square\varphi = 0$ and ψ its time derivative, namely

$$\varphi_\pm = (\varphi \pm i\omega^{-1}\psi)/2 . \qquad (3.2)$$

One can then rewrite Q_0 as

$$Q_0(t, \varphi, \psi) = 2 \sum_\pm \| (x\omega \pm it\nabla)\varphi_\pm \|_2^2 \qquad (3.3)$$

$$= 2 \sum_\pm \| x\omega \exp(\pm i\omega t)\varphi_\pm \|_2^2 \qquad (3.4)$$

where we have used the identity

$$\exp(\mp i\omega t) \, x \, \exp(\pm i\omega t) = x \pm i\, t\, \omega^{-1}\nabla . \qquad (3.5)$$

We can then prove the following estimate.

__Proposition 3.1__ Let $n \geq 2$, $2 < \ell \leq \ell_1$, and $(\varphi, \psi) \in \Sigma$. Then φ and $\omega^{-1}\psi$ belong to L^ℓ and satisfy the estimates

$$\| \varphi \|_\ell, \ \| \omega^{-1}\psi \|_\ell \leq C\, t^{-\gamma(\ell)} E_0(\varphi,\psi)^{\alpha(\ell)/2} Q_0(t, \varphi, \psi)^{(1-\alpha(\ell))/2} \qquad (3.6)$$

where $E_0(\varphi,\psi) = \| \psi \|_2^2 + \| \nabla\varphi \|_2^2$.

__Proof.__ We express φ and ψ in terms of $\varphi_\pm$ by using (3.2). We then estimate $\varphi_\pm$ in L^ℓ by Lemma 1.1 with $\alpha(\ell) = 1 + \delta(s)$ and obtain

$$\| \varphi_\pm \|_\ell \leq C\, t^{-\gamma(\ell)} \| \omega\, e^{\pm i\omega t}\varphi_\pm \|_s$$

$$\leq C\, t^{-\gamma(\ell)} \| \omega\, e^{\pm i\omega t}\varphi_\pm \|_2^{\alpha(\ell)} \| x\omega\, e^{\pm i\omega t}\varphi_\pm \|_2^{1-\alpha(\ell)} \qquad (3.7)$$

where the second inequality is obtained by applying the Hölder inequality in the form

$$\| \chi \|_s \leq \|(a^2 + x^2)^{-1/2}\|_m \|(a^2 + x^2)^{1/2}\chi\|_2$$

with $1/s = 1/m + 1/2$ and optimizing with respect to a. The result then follows from (3.4), (3.7) and the fact that

$$E_0(\varphi,\psi) = 2 \sum_{\pm} \|\omega\varphi_\pm\|_2^2 .$$

Q.E.D.

Although Proposition 3.1 is a result of the type we are looking for, it has some shortcomings. The limiting case $\ell = 2$ is excluded by the use of the Hölder inequality, whereas (3.6) for $\ell = 2$ and $n \geq 3$ is simply the Hardy inequality. Furthermore the upper limit $\ell \leq \ell_1$ is not optimal, as one expects the result to hold for all $\ell \leq 2^*$ ($\ell < \infty$ for n = 2). The upper limit $\ell = 2^*$ for $n \geq 3$ is sharp however, as shown by the following argument : for $\varphi = \varphi_+$, (3.7) takes the form

$$\| \varphi \|_\ell \leq c \, t^{-\gamma(\ell)} \|\omega\varphi\|_2^{\alpha(\ell)} \|(x\omega + it\nabla)\varphi\|_2^{1-\alpha(\ell)}$$

$$= c \, t^{1-\delta(\ell)} \|\omega\varphi\|_2^{\alpha(\ell)} \|(xt^{-1}\omega + i\nabla)\varphi\|_2^{1-\alpha(\ell)}$$

which for smooth and rapidly decaying φ implies $\| \varphi \|_\ell = 0$ and therefore $\varphi = 0$ if $\delta(\ell) > 1$, or equivalently $\ell > 2^*$, by letting t tend to infinity.

The second method will remedy the previous defects. We restrict our attention to the case $n \geq 3$ and refer to [8] for a detailed treatment of the case n = 2 which has additional complications in the form of logarithmic factors in the estimates. Since we want to estimate φ in terms of Q_0 which contains another function ψ independent of φ, we lose nothing by first minimizing Q_0 as a function of ψ. We introduce an auxiliary radial function $h \in \mathcal{C}((0, \infty); \mathbb{R})$ and we use the notation $\hat{x} = x/r$, $\theta = (t^2+r^2)^{1/2}$.

Lemma 3.1 Let $n \geq 3$, let $h \in \mathcal{C}^1((0, \infty); \mathbb{R})$ with h and rdh/dr in $L^\infty(\mathbb{R}^+)$ and let $\varphi \in \Sigma_1$. Then

$$Q_m(t,\varphi) \equiv \underset{\psi}{\text{Inf}} \; Q_0(t,\varphi,\psi) = \| \theta r^{-1} L \varphi\|_2^2$$

$$+ \|\theta^{-1}\{(t^2-r^2) \hat{x}.\nabla\varphi - (n-1-\theta^2 r^{-2} h)r\varphi\}\|_2^2$$

$$+ < \varphi, \{\theta^2 r^{-2} h(n-2-h) + (t^2 r^{-2}-1) \, rdh/dr\} \varphi > . \qquad (3.8)$$

We refer to [8] for the proof of Lemma 3.1, which is an elementary computation since $Q_0(t, \varphi, \psi)$ is a quadratic function of ψ. That lemma provides a control in the L^2-norm sense of (i) the angular derivatives of φ with an extra factor θ/r, (ii) the radial derivative of φ with a factor which vanishes on the light cone, since

$$\theta^{-1}|t^2-r^2|\, \hat{x}.\nabla\varphi \sim |t-r|\, d\varphi/dr$$

(by choosing $h = (n-1)r^2/\theta^2$), and (iii) φ itself with an extra factor θ/r (by choosing $h = (n-2)/2$). The last result is a variant of the Hardy inequality.

We can now state the main result of this section, which improves over Proposition 3.1.

Proposition 3.2 Let $n \geq 3$, let $2 \leq \ell \leq 2^*$, let $\tilde{\ell} = \text{Max}(\ell, 2n/(n-1))$, let $a \geq 0$, let $t > 0$ and $\varphi \in \Sigma_1$. Then φ satisfies the estimate

$$\|(\theta r^{-1})^{1-\delta(\ell)}\, \theta^{\gamma(\ell)}\, (\theta^{-1}|t^2-r^2|+ a)^{\alpha(\ell)}\, \varphi\|_\ell \leq C_0^{\delta(\ell)}$$

$$\times\ \|\theta r^{-1}\varphi\|_2^{1-\delta(\ell)}\{\|\theta r^{-1} L\varphi\|_2^2 + \|\theta r^{-1}\varphi\|_2^2\}^{\gamma(\ell)/2}$$

$$\times\ \{\|\theta^{-1}|t^2-r^2|\,(\frac{d\varphi}{dr} + \frac{n}{\tilde{\ell}r}\varphi)\|_2^2 + a^2\|\frac{d\varphi}{dr}\|_2^2\}^{\alpha(\ell)/2}$$

$$\leq C_1\, Q_m(t,\varphi)^{(1-\alpha(\ell))/2}(Q_m(t,\varphi) + a^2\|\frac{d\varphi}{dr}\|_2^2)^{\alpha(\ell)/2} \quad (3.9)$$

where the constants C_0 and C_1 depend only on n.

The proof is a complicated variant of the elementary proof [1] of the usual Sobolev inequality $\|u\|_{2^*} \leq C\|\nabla u\|_2$. In the latter one starts with the estimates

$$|v(x)| \leq (1/2) \int dx_j' |\partial_j v(x_1 \ldots, x_j', \ldots, x_n)|$$

for $1 \leq j \leq n$, one applies Hölder's inequality repeatedly, one substitutes $v = |u|^p$ for a suitable p, and one estimates the derivatives of u in L^2. Here we proceed similarly, but we use angular and radial instead of cartesian coordinates, and we introduce suitable weight factors in the starting point inequalities so as to reconstruct the weight factors occuring in $Q_m(t,\varphi)$ at the end of the computation. Eventually the angular derivatives of φ are controlled by the term $(\theta/r)L\varphi$ in Q_m, and the radial derivative is controlled by the radial term in Q_m

(which yields (3.9) with a = 0) or is left untouched (which gives the term with a in (3.9)). We refer to [8] for the details.

The estimate (3.9) for φ is stronger than the estimate (3.6) of Proposition 3.1 in every respect. In fact, the allowed values of ℓ range over the interval $[2, 2^*]$ for (3.9) instead of the interval $(2, \ell_1]$ for (3.6). In the left-hand side of (3.9), the L^ℓ-norm is improved by various additional factors : the factor $(\theta r^{-1})^{1-\delta(\ell)}$ yields the Hardy inequality, the factor $\theta^{\gamma(\ell)}$ yields the time decay $t^{-\gamma(\ell)}$ and an additional decrease at infinity in space; finally the factor $(\theta^{-1}|t^2-r^2|)^{\alpha(\ell)}$ yields an additional decay away from the light cone.

It is interesting to remark that the norm that appears in the left-hand side of (3.9) has some similarity with the norm used in [15] for n = 3 to prove the existence of global solutions for small data or equivalently the existence of dispersive solutions of the equation (1.1) down to the optimal values $p_0(3) = 1 + \sqrt{2}$, namely (see (4.9a) of [15])

$$\| (t+r) \, (1+ |t-r|)^{p-2} \, \varphi \|_\infty \ .$$

We finally remark that the estimates of Proposition 3.2 still hold if φ is a scalar field minimally coupled with an external Yang-Mills field, with the ordinary derivatives replaced by covariant derivatives. That property follows from the fact that for any function v with values in the space of Yang-Mills potentials or in the space relevant for the coexisting scalar fields, the following inequality holds

$$\partial_\mu |v| \le |D_\mu v| \ ,$$

where $|.|$ denotes the norm in the relevant space, and D_μ the covariant derivative corresponding to ∂_μ (see [5], especially the Appendix).

4. TIME DECAY

In this section, we apply the previous results to the derivation of time decay properties of the solutions of the equation (1.1). Those decay properties will result in a straightforward way from combining the estimate of Proposition 3.2 (for n ≥ 3) and its analogue for n = 2 with boundedness properties of $Q_0(t, \varphi, \dot{\varphi})$, and we concentrate our attention on the derivation of the latter. We assume throughout this section that $(\varphi_0, \psi_0) \in \Sigma$ and that f satisfies the assumptions (A1) and (A2). We need in addition a repulsivity condition (A3) :

(A3) f satisfies the inequalities

$$0 \leq (p_1+1) \; V(z) \leq 2 \; \text{Re} \; \bar{z} \; f(z) \tag{4.1}$$

for all $z \in \mathbb{C}$. (Equivalently, $V(R) = R^{p_1+1} v(R)$ where v is a non negative non decreasing function from $\mathbb{R}^+$ to $\mathbb{R}$).

In the case of a single power interaction (1.2) with $p = p_1$, (A3) simply says that $\lambda \geq 0$. Note also that (A3) implies (A2 b, c).

We first consider the easy case where $p_1 + 1 \geq \ell_1$.

<u>Proposition 4.1</u> Let $n \geq 2$, let f satisfies (A1), (A2) and (A3) with $p_1+1 \geq \ell_1$. Let $(\varphi_0, \psi_0) \in \Sigma$, and let $(\varphi, \dot{\varphi})$ be the solution of (1.1) in $\mathscr{C}(\mathbb{R}, \Sigma)$ with initial data (φ_0, ψ_0) at time zero, as described in Proposition 2.2. Then $Q_0(t, \varphi, \dot{\varphi})$ is uniformly bounded in time :

$$Q_0(t, \varphi, \dot{\varphi}) \leq Q(t, \varphi, \dot{\varphi}) \leq Q(0, \varphi_0, \psi_0)$$

$$= \|x \; \psi_0\|_2^2 + \|r \nabla \; \varphi_0\|_2^2 + \int dx \; r^2 \; v(\varphi_0) \; . \tag{4.2}$$

<u>Proof.</u> The result follows immediately from the fact that under the assumption (A3), one has $W(\varphi) \leq 0$ and $t \; R(t,\varphi) \leq 0$ (See (1.10)-(1.11)).

Q.E.D.

We next turn to the difficult case $p_1+1 \leq \ell_1$. We assume nevertheless that $p_1 \geq 1 + 4/n$, a condition which is satisfied by the values of p, $p_1(n)$ and $p_2(n)$, for all n. That condition also makes the assumption $\int dx \; r^2 \; v(\varphi_0)$ in Proposition 2.2 unnecessary. We need a slightly stronger version of the repulsivity condition (A3) :

(A3') f satisfies the inequalities

$$C|z|^{p_1+1} \leq (p_1+1) \; V(z) \leq 2 \; \text{Re} \; \bar{z} \; f(z) \tag{4.3}$$

for some $C > 0$ and all $z \in \mathbb{C}$. (Equivalently, if $V(R) = R^{p_1+1} v(R)$ one assumes in addition that $v(0) > 0$).

We follow the method sketched in the introduction. The first task consists in extracting some preliminary decay from the conservation law (See (1.29)) .

<u>Lemma 4.1</u> Let n ≥ 2, let f satisfy (A1), (A2) and (A3') and let

$$\mu \equiv n+1 - (p_1+1)(n-1)/2 \geq 0 \tag{4.4}$$

Let $(\varphi_0, \psi_0) \in \Sigma$, let $t_0 \in \mathbb{R}$ and let $(\varphi, \dot{\varphi})$ be the solution of (1.1) in $\mathscr{C}(\mathbb{R}, \Sigma)$ with initial data (φ_0, ψ_0) at time t_0. Then

$$m(t) \equiv Q(t, \varphi, \dot{\varphi}) + E(\varphi, \dot{\varphi}) \leq m(0)(1+t^2)^\mu , \tag{4.5}$$

$$\int dx\, V(\varphi) \leq m(0)(1+t^2)^{\mu-1} , \tag{4.6}$$

$$\| \varphi(t) \|_{p_1+1} \leq C\, m(0)^{1/(p_1+1)} (1+|t|)^{1-2\delta(p_1+1)} . \tag{4.7}$$

<u>Proof.</u> From the conservation law and from (A3), we obtain the integral form of the differential inequality

$$dm/dt = 2t \int dx\, W(\varphi) \leq 2\mu t \int dx\, V(\varphi)$$
$$\leq 2\mu t(1+t^2)^{-1} m(t) \tag{4.8}$$

which implies (4.5) and therefore (4.6). Finally (4.7) follows from (4.6) and (A3') .

Q.E.D.

In order to estimate $Q_0(t, \varphi, \dot{\varphi})$, we first rewrite it by using (1.12) as

$$Q_0(t, \varphi, \dot{\varphi}) = \|M \varphi\|_2^2 + \|L \varphi\|_2^2 + \|D \varphi\|_2^2 = \sum_A \|\phi_A\|_2^2 \tag{4.9}$$

where in addition to the angular momentum $L = x \times \nabla$, we have introduced the operators

$$M = t\nabla + x(d/dt) \tag{4.10}$$

$$D = x \cdot \nabla + t(d/dt) + n-1 , \tag{4.11}$$

and the sum in the last member of (4.9) runs over A = L, M, D, with $\phi_A = A\varphi$. The operators L, M and D are the infinitesimal generators of space rotations, of pure Lorentz transformations, and space-time dilations respectively. The expression (4.9) is the analogue in the NLW case of the expression $Q_{0S} = (1/2) \|J\varphi\|_2^2$ in the NLS case (see (1.8)$_S$ and (1.21)). Note that in that case, J is the infinitesimal generator of pure Galilei transformations. From the transformation properties of the free wave equation, or more simply from a direct computation, it follows that

$$[\Box, A] = 0 \text{ for } A = L, M \; ; \quad \Box D = (D+2) \Box \quad , \tag{4.12}$$

so that the functions ϕ_A satisfy the equations

$$\begin{cases} \Box \, \phi_A + A f = 0 \quad \text{for } A = L, M , \\[4mm] \Box \, \phi_D + (D+2)f = 0 . \end{cases} \tag{4.13}$$

The interaction term in (4.13) can be rewritten as

$$Af = \phi_A \frac{\partial f}{\partial \varphi} + \overline{\phi}_A \frac{\partial f}{\partial \overline{\varphi}} \equiv \phi_A \, f'(\varphi) \qquad \text{for } A = L, M \tag{4.14}$$

$$(D+2)f = \phi_D \, f'(\varphi) + (n+1)f - (n-1) \varphi \, f'(\varphi).$$

The Cauchy problem for (1.1) with initial data (φ_0, ψ_0) at time zero can be rewritten in the form of the integral equation

$$\varphi(t) = \varphi^{(0)}(t) - \int_0^t d\tau \, K(t-\tau) f(\varphi(\tau)) \tag{4.15}$$

where

$$\varphi^{(0)}(t) = K(t) \varphi_0 + K(t) \psi_0 . \tag{4.16}$$

Similarly, the functions $\phi_A(t)$ satisfy the integral equations

$$\phi_A(t) = \phi_A^{(0)}(t) - \int_0^t d\tau \, K(t-\tau) \, \phi_A(\tau) \, f'(\varphi(\tau)) \qquad \text{for } A = L, M$$

and a similar equation for ϕ_D, with the functions $\phi_A^{(0)}$ expressible in terms of (φ_0, ψ_0). One sees easily that $\phi_A^{(0)}$ is bounded in L^2 uniformly in time (with an additional Log t factor for $\phi_M^{(0)}$ if $n = 2$) and it remains to estimate ϕ_A in L^2 by using the integral equations. The function $f'(\varphi)$ in the integrand is estimated by the use of the estimates (4.5)-(4.7) and possibly additional estimates obtained therefrom by substituting them in the integral equation (4.15) for φ . In space dimension $n \geq 3$, the method requires the use of homogeneous Besov spaces [2] and of estimates of the operator $K(t)$ acting between such spaces. It is reasonably simple in dimensions 3 and 4. The case of higher dimensions is more complicated and has not been worked out in details.

Here we only state the final results for n = 2, 3, 4, and refer to [8] and [9] for the details of the proofs in dimensions n = 2 and n = 3, 4 respectively. In the latter case, one needs a slight reinforcement of the assumption (A1), in the following form

(A1') $f \in \mathcal{C}^2(\mathbb{C}, \mathbb{C})$, $f(0) = f'(0) = 0$, and f satisfies the estimate

$$|f''(z)| = \text{Max}(|\partial^2 f/\partial z^2|, |\partial^2 f/\partial z \, \partial\bar{z}|, |\partial^2 f/\partial\bar{z}^2|) \leq C(|z|^{p_1 - 2} + |z|^{p_2 - 2})$$

for some p_1, p_2 with $1 \leq p_1 \leq p_2 < 1 + 4/(n-2)$ and all $z \in \mathbb{C}$.

<u>Proposition 4.2</u> If n = 2, let f satisfy (A1), (A2) and (A3') with $2 + \sqrt{5} \equiv p_2(2) < p_1 (\leq 5)$. If n = 3, 4, let f satisfy (A1'), (A2) and (A3') with $p_2(n) < p_1 (\leq 1 + 4/(n-1))$ and $p_1 \leq p_2 \leq 1 + 4n \, [(n-2)(n+1)]^{-1}$. Let $(\varphi_0, \psi_0) \in \Sigma$, let $t_0 \in \mathbb{R}$ and let $(\varphi, \dot{\varphi}) \in \mathcal{C}(\mathbb{R}, \Sigma)$ be the solution of (1.1) with initial data (φ_0, ψ_0) at time t_0. Then

$$Q_0(t, \varphi, \dot{\varphi}) \leq C(1 + \text{Log}_+|t|)^2 \qquad \text{if } n = 2 \quad,$$

$$Q_0(t, \varphi, \dot{\varphi}) \leq C \qquad \text{if } n = 3, 4 \quad.$$

The time decay properties of the solutions of (1.1) follow from Proposition 3.2, its analogue for n = 2, and Propositions 4.1 and 4.2. In particular all solutions of (1.1) with initial data in Σ satisfy the optimal decay (1.15) for $2 \leq \ell \leq 2^*$ under the assumptions of Proposition 4.1 for $n \geq 3$ and of Proposition 4.2 for n = 3, 4.

REFERENCES

[1] R.A. Adams, Sobolev spaces, Academic Press, New York 1975.

[2] J. Bergh, J. Löfström, Interpolation Spaces, Springer, Berlin 1976.

[3] J. Ginibre, G. Velo, in Non Linear PDE and their applications, Collège de France Séminar, Vol. II, H. Brézis, J.L. Lions Eds., Pitman, London (1982) p. 155-199.

[4] J. Ginibre, G. Velo, J. Funt. Anal. <u>32</u> (1979), p. 33-71.

[5] J. Ginibre, G. Velo, Commun. Math. Phys. <u>82</u> (1981), p. 1-28.

[6] J. Ginibre, G. Velo, unpublished.

[7] J. Ginibre, G. Velo, J. Math. Pur. Appl. <u>64</u> (1985), p. 363-401.

[8] J. Ginibre, G. Velo, Ann. IHP (Phys. Théor.) <u>47</u> (1987), p. 221-261.

[9] J. Ginibre, G. Velo, Ann. IHP (Phys. Théor.) <u>47</u> (1987), p. 263-276.

[10] J. Ginibre, G. Velo, The Cauchy problem for the non linear
     Klein Gordon equation, preceding lecture in these proceedings.

[11] R.T. Glassey, Math. Z. 177 (1981), p. 323-340.

[12] R.T. Glassey, Math. Z. 178 (1981), p. 233-261.

[13] R.T. Glassey, H. Pecher, Manuscripta Math. 38 (1982), p. 387-400.

[14] N. Hayashi, Y. Tsutsumi, Remarks on the scattering problem for non
     linear Schrödinger equations, in Proceedings of UAB Conference on
     Differential Equations and Mathematical Physics, Springer, Berlin,
     1986.

[15] F. John, Manuscripta Math. 28 (1979), p. 235-268.

[16] T. Kato, Ann. IHP (Phys. Théor.) 46 (1987), p. 113-129.

[17] B. Marshall, W. Strauss, S. Wainger, J. Math. Pure Appl. 59 (1980),
     p. 417-440.

[18] K. Mochizuki, T. Motai, J. Math. Kyoto Univ. 25 (1985), p. 703-715.

[19] K. Mochizuki, T. Motai, The scattering theory for the non linear
     wave equation with small data II, preprint (1986).

[20] H. Pecher, Math. Z. 150 (1976), p. 159-183.

[21] H. Pecher, Scattering for semilinear wave equations with small
     data in three space dimensions, Wuppertal preprint (1987).

[22] T. Sideris, J. Diff. Eq. 52 (1984), p. 378-406.

[23] T. Sideris, private communication.

[24] W. Strauss, J. Funct. Anal. 2 (1968), p. 409-457.

[25] W. Strauss, J. Funct. Anal. 41 (1981), p. 110-133.

[26] R. Strichartz, Trans. Amer. Math. Soc. 148 (1970), p. 461-470.

[27] Y. Tsutsumi, Thesis, University of Tokyo, 1985.

[28] Y. Tsutsumi, Ann. IHP (Phys. Théor.) 43 (1985), p. 321-347.

[29] W. Von Wahl, J. Funct. Anal. 9 (1972), p. 490-495.

Energy Forms and White Noise Analysis

T. Hida
Department of Mathematics
Nagoya University, Japan

J. Potthoff
Fachbereich Mathematik
Technische Universität Berlin, FRG

L. Streit
BiBoS
Universität Bielefeld, FRG

In this article we give a short introduction to white noise analysis, with special emphasis to generalized and positive generalized white noise functionals and the causal calculus. These will be the main ingredients for the discussion of Dirichlet and energy forms in this framework as examples of application [8,9]. For more details on white noise analysis see e.g. [6,7,11,12] and literature quoted there.

1. White Noise

For simplicity we shall work in this article mostly with white noise with one dimensional time parameter. The generalization of notions and statements to higher dimensional "time" will be obvious. Thus we consider the Gel'fand triple

$$S'(\mathbb{R}) \supseteq L^2(\mathbb{R}, dt) \supseteq S(\mathbb{R}) \qquad (1.1)$$

which by Minlos' theorem [3,5] induces a Gaussian measure $d\mu$ on the σ-algebra $\mathcal{B}$ over $S'(\mathbb{R})$ generated by the cylinder sets such that for all $\xi \in S(\mathbb{R})$

$$\int_{S'(\mathbb{R})} e^{i\langle x, \xi \rangle} d\mu(x) = e^{-1/2(\xi, \xi)} \qquad (1.2)$$

where we denoted the pairing of $S'(\mathbb{R})$ and $S(\mathbb{R})$ by $\langle \cdot, \cdot \rangle$ and the $L^2(\mathbb{R}, dt)$ scalar product by $(\cdot, \cdot)$. In fact one can prove that $d\mu$ is supported by the Sobolev space $S_{-\alpha}(\mathbb{R}, dt)$, $\alpha > 1/2$, which is defined as the completion of $S(\mathbb{R})$ under the norm

$$\| f \|_{-\alpha} = \| H^{-\alpha} f \| \tag{1.3}$$

where the last norm is the one of $L^2(\mathbb{R}, dt)$ and H is the Hamiltonian of the harmonic oscillator: $H = -d^2/dt^2 + t^2 + 1$ (we have added 1 to the usual Hamiltonian for later purposes).

For $p \geq 1$, we shall denote the Banach space $L^p(S'(\mathbb{R}), \mathcal{B}, d\mu)$ by (L^p).

It is well-known (e.g. [5]) that (L^2) is isometric to the symmetric Fock space over $L^2(\mathbb{R}, dt)$ (complexified)

$$(L^2) \simeq \overset{\infty}{\underset{n=0}{\oplus}} L^2(\widehat{\mathbb{R}^n, n! \, d^n t}) \tag{1.4}$$

where $\hat{\cdot}$ denotes symmetrization and if $n = 0$ we mean $\mathbb{C}$. Therefore (L^2) admits a direct sum decomposition $(L^2) = \overset{\infty}{\underset{n=0}{\oplus}} H^{(n)}$, and $H^{(n)}$ is called the n-th homogeneous chaos. Elements $\varphi^{(n)}$ in $H^{(n)}$ can be visualized in the following way

$$\varphi^{(n)}(x) = \int_{\mathbb{R}^n} f^{(n)}(t) :x^{\otimes n}: (t) d^n t, \quad x \in S'(\mathbb{R}) \tag{1.5}$$

where $f^{(n)} \in L^2(\widehat{\mathbb{R}^n, n! \, d^n t})$ and $:\;:$ denotes a "Wick ordered product". It is defined recursively by

$$:x^0: \quad = 1$$

$$:x: (t_1) = x(t_1), \quad t_1 \in \mathbb{R} \tag{1.6}$$

$$:x^{\otimes n}: (t) = x(t_n) :x^{\otimes (n-1)}: (t_1, \ldots, t_{n-1})$$

$$- \sum_{k=1}^{n-1} \delta(t_n - t_k) :x^{\otimes (n-2)}: (t_1, \ldots, \not{t}_k, \ldots, t_{n-1})$$

$(t = (t_1,...,t_n))$. Then it takes a moment's thought to see that for $f^{(n)} \in S(\mathbb{R}^n)$ (1.5) is well-defined, if one interpretes the integral on the right as dual pairing. For $f^{(n)} \in L^2(\hat{\mathbb{R}}^n)$ (1.5) has to be interpreted as the (L^2)-limit of expressions like (1.5) with $f^{(n)}$ approximated by functions in $S(\mathbb{R}^n)$. In terms of $\varphi = \sum_n \varphi^{(n)} \in (L^2)$, $\varphi^{(n)}$ as in (1.5), the isometry (1.4) reads

$$\|\varphi\|^2_{(L^2)} = \sum_{n=0}^{\infty} n! \, \|f^{(n)}\|^2_{L^2(\mathbb{R}^n, d^n t)} \tag{1.7}$$

Let us now introduce a useful transformation on (L^2) [11]:

$$(S\varphi)(\xi) := \int \varphi(x+\xi) d\mu(x), \quad \xi \in S(\mathbb{R}) \tag{1.8}$$

One readily checks that $\varphi^{(n)}$ in (1.5) transforms as:

$$(S\varphi^{(n)})(\xi) = \int_{\mathbb{R}^n} f^{(n)}(t) \xi^{\otimes n}(t) d^n t \tag{1.9}$$

which we may view as the evaluation of the tensor $f^{(n)} \in (L^2(\mathbb{R}, dt))^{\hat{\otimes} n}$ on $(\xi,...,\xi) \in (L^2(\mathbb{R}, dt))^{\otimes n}$; in other words, we identify $S\varphi^{(n)}$ in a unique way with an element in $(L^2(\mathbb{R}, dt))^{\hat{\otimes} n}$ and therefore the range of S is the symmetric Fock space over $L^2(\mathbb{R}, dt)$: S implements the isometry (1.4).

We conclude this subsection by giving attention to a special element in (L^2), in fact in $H^{(1)}$, namely $B(t;x) = <x, 1_{[0,t)}>$, $t > 0$, (understood as an (L^2)-limit, s.a.) which is a version of Brownian motion and due to this relation we interpret $\dot{B}(t;x) \equiv x(t)$ as the time derivative of Brownian motion.

2. Generalized Functionals

There are many motivations for the introduction of generalized functionals of white noise, but here we confine ourselves to mention that in section 4 we shall construct energy forms based on certain positive generalized functionals. For other motivations we refer the reader to the literature quoted in the references. Also, it should be pointed out that the type and construction of such spaces of generalized

functionals - like in finite dimensions - will vary with the application one has in mind. For our purposes it is convenient to use a construction based on second quantized operators (cf. also [11,15,19]).

Consider again the symmetric Fock space over $L^2(\mathbb{R}, dt)$:

$$\hat{F} = \bigoplus_{n=0}^{\infty} (L^2(\mathbb{R}, dt))^{\hat{\otimes}n} \qquad (2.1)$$

(which we identify with the r.h.s. of (1.4)). If A is a linear, closable operator we may define it second quantization $\Gamma(A)$ on $\hat{F}$ by
setting

$$\Gamma(A)\Big|_{(L^2(\mathbb{R}, dt))^{\hat{\otimes}n}} = A^{\otimes n} \qquad (2.2)$$

and extending (2.2) linearly [16]. We choose $A = H$, H being the
Hamiltonian of the harmonic oscillator (cf. section 1). Note that

$$\Gamma(H)\Big|_{(L^2(\mathbb{R}, dt))^{\hat{\otimes}n}} \geq 2^n \qquad (2.3)$$

The isometry (1.4) defines now a unitary image of $\Gamma(H)$ on (L^2) and
we denote it by the same symbol for simplicity. Moreover we set for
$p \in \mathbb{N}_0$

$$(S_p) := \mathcal{D}(\Gamma(H^p)) \subset (L^2) \qquad (2.4)$$

$\mathcal{D}(\Gamma(H^p))$ denoting the completed domain of $\Gamma(H^p)$. Clearly (S_p) is
a Hilbert subspace of (L^2). Also it is not hard to see that the sys-
tem of norms $(\|\cdot\|_{2,p}; p \in \mathbb{N}_0)$ of the spaces (S_p) is a compatible
system (see e.g. [3]). Let us define the space

$$(S) := \bigcap_{p \in \mathbb{N}_0} (S_p) \qquad (2.5)$$

as the projective limit of the family $((S_p); p \in \mathbb{N}_0)$. (S) is a
countably Hilbert nuclear space [3,11]. Its dual is given by [3]

$$(S)^* = \bigcup_{p \in \mathbb{N}_0} (S_{-p}) \qquad (2.6)$$

where (S_{-p}) is the dual of (S_p). We have therefore obtained a nuclear rigging

$$(S)^* \supseteq (L^2) \supseteq (S) \tag{2.7}$$

and it is easy to check that (S) is dense in (L^2) (cf. [8] for an argument). White noise functionals in (S) are called <u>testfunctionals</u>, while those in $(S)^*$ are called <u>generalized functionals</u>. The dual pairing between $(S)^*$ and (S) will also be denoted by $<\cdot,\cdot>$.

A very useful property of (S) is [8,11]

<u>Lemma 2.1</u>:

(S) is an algebra.

In order to give the reader an intuitive idea about the space (S), we remark that in the representation (1.5) of $\varphi^{(n)}$ belonging to $(S) \cap H^{(n)}$ $f^{(n)}$ is a member of $\hat{S}(\mathbb{R}^n)$ (symmetric Schwartz space) and $\|\varphi^{(n)}\|_{(L^2)}$ falls off in n faster than any exponential.

From the last remark it should be clear that there are elements in $(S)^*$ which correspond formally to functionals like (1.5) with $f^{(n)} \in \overset{\wedge}{S'}(\mathbb{R}^n)$, e.g. $:x(t)^n:$ and the like.

Another class of functionals in $(S)^*$, so-called <u>Gauss kernels</u>, will be described next. This class will later provide the first examples of singular measures which yield energy forms.

Formally speaking we want to construct elements like

$$\Phi_{un}(x) = \exp(-1/2 <x,Kx>) \tag{2.8}$$

in $(S)^*$, where K is some operator on $L^2(\mathbb{R}, dt)$. In [18] (2.8) had been studied for K of Hilbert-Schmidt type with the following result: if $K > -1$, $\|K(1+K)^{-1}\|_{H.S.} < 1$ and K is trace class, then (2.8) exists (as it stands) in (L^2). If the trace class condition is dropped, Φ_{un} belongs after a multiplicative renormalization ("division by the expectation") to (L^2): we denote this element by

$$\Phi(x) = N \exp(-1/2 <x,Kx>) \tag{2.9}$$

For more general linear selfadjoint operators K on $L^2(\mathbb{R}, dt)$ we proved in [9] the following

Lemma 2.2:

Assume that $K > -1$ and that for some $p \in \mathbb{N}$

$$\text{(i)} \qquad \| H^{-p} K (1+K)^{-1} H^{-p} \|_{H.S.} < 1 \qquad\qquad (2.10.a)$$

$$\text{(ii)} \qquad \| H^{-p} (1+K)^{-2} H^{-p} \|_{H.S.} < \infty \qquad\qquad (2.10.b)$$

Then there is an element Φ in $(S)^*$ (in fact in (S_{-p})), formally denoted as in (2.9), with characteristic functional

$$<\Phi, \exp i<\cdot,\xi>> = \exp(-1/2(\xi,(1+K)^{-1}\xi)) \qquad\qquad (2.11)$$

$(\xi \in S(\mathbb{R}))$.

Remark: Without condition (2.10.b) this has been shown in [9] for K with $K \geq -1+\varepsilon$, $\varepsilon > 0$. The above lemma can be proved on the basis of this result by a simple limiting argument, which makes use of (2.10.b).

Gauss kernels have an important property: they are positive in the sense that they map every positive (a.c.) element in (S) into a positive number. Elements in $(S)^*$ with this property will be called from now on positive generalized functionals. A useful theorem about positive generalized functionals is found in a paper by Yokoi [19]:

Theorem 2.3:

If $\Phi \in (S)^*$ is positive, then there exists a unique finite positive measure ν_Φ on B so that for every $F \in (S)$

$$<\Phi,F> = \int \tilde{F}(x) d\nu_\Phi(x) \qquad\qquad (2.12)$$

where $\tilde{F}$ is the continuous version of F.

3. White Noise Calculus

This subsection will be rather sketchy, since by now the white noise calculus can be considered as being well-established. For further details we refer the interested reader to [5,9,11,12] and references quoted there.

The idea is to regard the generalized random variable $x(t)$, i.e. white noise "at time t", as a continuously indexed system of coordinates and to build a differential calculus with these coordinates. Thus we introduce a differential operator $\partial_t = \partial/\partial x(t)$ as follows

Definition 3.1:

On (S) we set

$$\partial_t := S^{-1} \frac{\delta}{\delta \xi(t)} S, \quad t \in \mathbb{R} \tag{3.1}$$

where $\delta/\delta\xi(t)$ is the Fréchet functional derivative, and S is the transformation (1.8).

With this definition one can check that ∂_t is indeed a derivation on (S) (i.e. it admits the Leibniz rule) and also one can prove a chain rule [14]. Of course the domain of definition of ∂_t can be largely extended. Here we confine ourselves to mention the following results

Lemma 3.2:

(i) ∂_t maps (S) continuously into itself

(ii) ∂_t is an annihilation operator: if we denote $H_+^{(n)} := (S) \cap H^{(n)}$ (cf. section 2), then

$$\partial_t: H_+^{(n)} \rightarrow H_+^{(n-1)} \tag{3.2}$$

The adjoint operator ∂_t^* of ∂_t is defined as that

$$<\partial_t^*\Phi,F> = <\Phi,\partial_t F> \tag{3.3}$$

for all $\Phi \in (S)^*$ and all $F \in (S)$.

Lemma 3.3:

(i) ∂_t^* maps $(S)^*$ continuously (in the weak-*-topology) into itself

(ii) ∂_t^* is a creation operator: denote $H_-^{(n)} := (H_-^{(n)})^*$, then

$$\partial_t^*: H_-^{(n)} \to H_-^{(n+1)} \tag{3.4}$$

(iii)   we have the canonical commutation relations

$$[\partial_t, \partial_s^*] = \delta(t-s) \tag{3.5.a}$$

$$[\partial_t, \partial_s] = [\partial_t^*, \partial_s^*] = 0 \tag{3.5.b}$$

(iv)   multiplication by $x(t)$, $t \in \mathbb{R}$,   is defined as a mapping from
      (S) into $(S)^*$ by

$$x(t) \cdot = \partial_t^* + \partial_t \tag{3.6}$$

Formula (3.6) is the starting point of a reformulation of stochastic integrals within the white noise calculus, which leads to important generalizations of the standard stochastic integrals, cf. [13].

4. Energy Forms

In this subsection we shall sketch how energy forms [2] can be handled within the framework of white noise analysis. For more details and references for other approaches of energy forms in infinite dimension we refer the reader to our paper [8]. A review about the finite dimensional case is given in [17].

Roughly speaking the main idea is to generalize the finite dimensional form

$$\varepsilon(f,g) := \int_{\mathbb{R}^n} (\nabla f)(x) \cdot (\nabla g)(x) \, d\nu(x) \tag{4.1}$$

where $d\nu$ is some positive σ-finite measure - the "ground-state measure" - to an infinite dimensional situation. If ε in (4.1) is closable, then it defines uniquely a selfadjoint positive operator $\hat{H}$ on

$L^2(\mathbb{R}^n, d\nu)$ [10]. This representation of quantum mechanics allows for constructions of Hamiltonians $\hat{H}$ which in the Schrödinger representation with $H = -\Delta + V$ would have very singular potentials V. Thus this strategy seems promising for quantum field theory.

We shall work in this section with white noise with d-dimensional time parameter, i.e. with the basic triplet

$$S'(\mathbb{R}^d) \supseteq L^2(\mathbb{R}^d, dt) \supseteq S(\mathbb{R}^d) \qquad (4.2)$$

and without further mentioning we shall use the notions introduced in the preceding sections in this context.

From how on let Φ be some positive generalized functional in $(S)^*$. By theorem 2.3 Φ defines a measure on $S'(\mathbb{R}^d)$ which we shall denote by $\Phi(x)d\mu(x)$, $d\mu$ being the white noise measure, although this measure is in general not absolutely continuous with respect to $d\mu$.

On the basis of lemma 2.1 and lemma 2.2 (cf. also the remark after lemma 2.1) it is not hard show that the following lemma holds.

Lemma 4.1:

If $F,G \in (S)$, then also $\nabla F \cdot \nabla G \in (S)$ with

$$\nabla F := (\partial_t F;\ t \in \mathbb{R}^d) \qquad (4.3)$$

and

$$\nabla F \cdot \nabla G = \int_{\mathbb{R}^d} dt\,(\partial_t F)(\partial_t G) \qquad (4.4)$$

Therefore we may define the following sesquilinear form on (S):

$$\varepsilon(F,G) := \langle \Phi, \nabla \bar{F} \cdot \nabla G \rangle \qquad (4.5)$$

$$= \int (\nabla \bar{F}(x)) \cdot (\nabla G(x)) \Phi(x) d\mu(x) \qquad (4.6)$$

Equality (4.6) follows from theorem 2.3 (upon identification of elements in (S) with their continuous representant). Since (S) is dense in $(L^2)_\Phi := L^2(S'(\mathbb{R}^d), \Phi d\mu)$, ε is a densely defined, symmetric form on $(L^2)_\Phi$.

Next arises the question of closability of the form ε in $(L^2)_\Phi$, which would guarantee the existence of a unique selfadjoint operator $\hat{H}$ on $(L^2)_\Phi$ with

$$\overline{\varepsilon}(F,G) = (F,\hat{H}G)_\Phi \tag{4.7}$$

for all $F,G \in \mathcal{D}(\hat{H}^{1/2}) = \mathcal{D}(\overline{\varepsilon})$, where $\overline{\varepsilon}$ is the closure of ε and $(\cdot,\cdot)_\Phi$ is the $(L^2)_\Phi$ scalar product, [10,p.322].

Definition 4.2:

Let $\Phi \in (S)^*$ be positive. Φ is called <u>admissible</u> iff its associated symmetric form ε (cf. (4.5)) is closable.

We can prove the following criterium [8,9]

Theorem 4.3:

If $\Phi \in (S)^*$, positive, is such that for every $t \in \mathbb{R}^d$ $\partial_t\Phi = B(t)\Phi$ with

$$\int B(t)\eta(t)dt \in (S) \tag{4.8}$$

for every $\eta \in S(\mathbb{R}^d)$, then Φ is admissible.

Example 4.4:

The Gauss kernels of section 2 are admissible, if the operator K (cf. (2.9)) on $L^2(\mathbb{R}^d, dt)$ is such that K maps $S(\mathbb{R}^d)$ into itself. Specially, one can choose $K = (-\Delta+m^2)^{1/2}$, $m^2 > 0$. As is shown in [9], this provides a representation of the free relativistic massive boson field in d space dimensions as a generalized white noise functional.

For a criterium in the case that the measure induced by Φ is absolutely continuous w.r.t. $d\mu$, we refer to our paper [8].

In [1] also non-Gaussian measures will be treated.

References

[1] S. Albeverio, T. Hida, J. Potthoff and L. Streit, work in preparation.

[2] M. Fukushima, Dirichlet forms and Markov processes, North-Holland Kodansha, 1980.

[3] I.M. Gel'fand and N.Ya. Vilenkin, Generalized functions, vol. 4, Academic press, New York, 1964.

[4] T. Hida, Analysis of Brownian functionals, Carleton Mathematical Lecture Notes, no.13, 1975, 2nd edition 1978.

[5] T. Hida, Brownian motion, Japanese ed. 1975, English translation 1980, Springer-Verlag, Applications of Math., vol.11.

[6] T. Hida, Brownian functionals and rotation group, Mathematics + Physics, ed. by L. Streit, vol.1 (1985), 167-194.

[7] T. Hida, H.-H. Kuo, J. Potthoff and L. Streit, White noise: an infinite dimensional calculus, monograph in preparation.

[8] T. Hida, J. Potthoff and L. Streit, Dirichlet forms and white noise analysis, to appear in Commun. Math. Phys.

[9] T. Hida, J. Potthoff and L. Streit, White noise analysis and applications, Mathematics + Physics, vol.3, ed. by L. Streit, World Scientific Publ. Co., Singapore, to appear.

[10] T. Kato, Perturbation theory for linear operators, Springer-Verlag, 1976.

[11] I. Kubo and S. Takenaka, Calculus on Gaussian white noise, I-IV, Proc. Japan Academy, A, Math. Sci., 56, (1980), 376-380, 411-416; 57, (1981), 433-437; 58, (1982), 186-189.

[12] H.-H. Kuo, Brownian functionals and applications, Acta Applicandae Math., 1, (1983), 175-188.

[13] H.-H. Kuo and A. Russek, White noise approach to stochastic integration, preprint (1986).

[14] J. Potthoff, White noise approach to Malliavin calculus, J. Functional Analysis, 71, (1987), 207-217.

[15] J. Potthoff, On positive generalized functionals, J. Functional Analysis, 74, (1987), 81-95.

[16] B. Simon, The $P(\phi)_2$ Euclidean (quantum) field theory, Princeton Univ. Press, 1974.

[17] L. Streit, Energy forms: Schrödinger theory, Processes, Phys. Reports, 77 (1980).

[18] L. Streit and T. Hida, Generalized Brownian functionals and the Feynman integral, Stochastic Processes and their Applications, 16, (1984), 55-69.

[19] Y. Yokoi, Positive generalized functionals, Preprint.

Principles of Solitary Wave Stability

Luis Vazquez

Dpto. de Fisica Teórica, Universidad Complutense

28040 - Madrid, Spain

1. Introduction

A lot of physical systems can be described by nonlinear differential equations
which admit solutions in the form of so-called solitary waves. By a solitary wave
we mean a localized wave which keeps its form or shape. Physically, a basic problem
is to understand the role of these localized nonlinear objects. One of the most important and natural requirements for solitary waves is the condition of stability.

For example, in particle physics classical nonlinear wave equations form a basis
for the construction of quantum objects. In all investigations the first order
approximation deals with the equations as if they were describing classical field
configurations, rather than quantum operator fields. In such a case the demand for
classical stability is motivated by the requirement that the corresponding quantum
state should be stable.[J]

Unfortunately the meaning of stability is not unique in the sense that there exist
a lot of different definitions (or notions) of stability used in the literature and
the information about the relations between these concepts of stability are very mea-
ger. The aim of this contribution is therefore to give a brief survey about the
stability problem and to report on recent progress in this field. Furthermore using
simple conditions we prove some relations between several concepts of solitary wave
stability.

In Section 2 we present a general framework for the stability of solitary waves.
We restrict ourselves to Hamiltonian systems since the Hamiltonian structure will
allow to extend all the stability methods developed in classical mechanics (i.e.
for systems having finite degrees of freedom) to systems with infinite degrees of
freedom.

In Section 3 we study the relations between several concepts of stability and
prove some simple theorems on them. In particular, we are interested in the relation
between energetic and nonlinear stability.

In Section 4 we report on recent numerical work concerning soliton stability in
nonintegrable systems.

2. Concepts of Stability

As pointed out in the introduction we shall consider Hamiltonian systems only. In the standard form they can be written as

(H)
$$\frac{du}{dt} = J E'(u)$$

on a function space (e.g. a Hilbert space) where E denotes the energy, E' its derivative 'with respect to u' and J is a skew-symmetric linear operator.

Typical examples for (H) on an infinite dimensional space which arise from classical field theory are nonlinear Klein-Gordon equations

(NLKG)
$$\psi_{tt} - \Delta\psi + m^2\psi - g(|\psi|^2)\psi = 0 \qquad \text{on } \mathbf{R}^N$$

nonlinear Schrödinger equations

(NLS)
$$i\phi_t + \Delta\phi + f(|\phi|^2)\phi = 0 \qquad \text{on } \mathbf{R}^N$$

and nonlinear Dirac equations

(NLD)
$$i\gamma^0\partial_t\psi + i\gamma^k\partial_k\psi - m\psi + \frac{\partial G(\overline{\psi}\Gamma\psi)}{\partial\overline{\psi}} = 0 \quad \text{on } \mathbf{R}^N$$

$$1 \leq k \leq N, \quad \Gamma = \text{element of the } \gamma\text{-algebra}.$$

An interesting property of all these equations is that they possess additional symmetries like translation invariance and (global) gauge invariance. These symmetries generate additional conserved quantities like momentum (coming from translation invariance) and charge (coming from gauge invariance). Therefore we may assume that (H) is invariant under certain group representations. We follow now a recent paper where a rather general stability theory has been presented [GSS]. We assume That (H) is invariant under a one-parameter group of operators U($\cdot$) (an extension to more dimensional abelian groups seems not to be difficult). Let Q be the conserved quantity associated to U($\cdot$). A solitary wave solution is a solution of the form

(2.1)
$$u(t) = U(\omega t)\phi_\omega$$

It may be viewed as a critical point of the energy E subject to constant Q. This leads to the definition of energetic stability.

Definition 1: A solution of (H) given by (2.1) is called energetically stable if its stationary part ϕ_ω is a local minimum of the energy E subject to constant Q.

A sufficient condition for energetic stability is that the linearized operator $H_\omega = E''(\phi_\omega) - \omega Q(\phi_\omega)$ is nonnegative. One should note that H_ω cannot be positive since the symmetry causes a nontrivial nullspace (see [GSS]).

Orbital (or nonlinear) stability is defined as follows.

Definition 2: The solution $U(\omega t)\phi_\omega$ is called orbitally stable if for any tubular neighborhood of $O_\omega = \{U(s)\phi_\omega, s \text{ real}\}$ there exists a neighborhood V of ϕ_ω such that all trajectories $u(t)$ of (H) which start in V remain in the given tube for all time. Otherwise we call $U(\omega t)\phi_\omega$ unstable.

In most cases one uses energetic stability to prove orbital stability by taking the energy as a Liapunov functional [Be] but energetic stability is not a necessary condition for orbital stability [JR,SV].

Another method to study the stability properties of solitary wave solutions is to linearize the system around a solitary wave $U(\omega t)\phi_\omega$. The linearized dynamics is then described by the equation

$$(\text{H}_{\text{lin}}) \qquad \frac{dw}{dt} = J H_\omega w$$

with $H_\omega = E''(\phi_\omega) - \omega Q''(\phi_\omega)$.

Definition 3: A solitary wave $U(\omega t)\phi_\omega$ is called linear dynamically stable if any solution of the linearized system remains bounded for $t \geq 0$.

Unfortunately this definition is not directly applicable since one has to solve the so called zero-mode problem [ML]: The nontrivial nullspace of the linearized operator in the case of symmetries will generate solutions which grow polynomial in time. Physically, however, these zero modes (or secular modes) have no meaning for the stability properties. Apart from some particular cases [W1] we do not have a rigorous result for the linear dynamical stability.

Physically, an interesting problem is to study the 'collision' of two solitary waves. If one interprets solitary waves as particles then the interaction between two solitary waves yields information on the 'stability' of the considered formation. If the solitary waves interact elastically then they have (at least asymptotically) the same shape as before. In such a case these solitary waves are called soliton. Mathematically the soliton property is related to the complete integrability of the corresponding field equation. Solitons have been found analytically by the inverse scattering transform, e.g. for the sine-Gordon equation

$$(\text{S-G}) \qquad \psi_{tt} - \psi_{xx} + \sin \psi = 0.$$

This method also permits an analytically study of soliton interactions.

Unfortunately nonintegrable systems, e.g. the ϕ^4-model

$$(\phi^4) \qquad \phi_{tt} - \phi_{xx} - \phi + \phi^3 = 0 ,$$

abound in physics and the powerful methods available for integrable systems do not apply to these models.

3. Relations between energetic linear and nonlinear stability

To discuss the relation between energetic and nonlinear stability we start with a rather general but simple theorem.

Theorem 3.1: Let ϕ_0 be a solution of (H) , i.e. $E'(u_0) = 0$ such that the linear operator $E''(\phi_0)$ satisfies

$$(3.1) \qquad <E''(\phi_0)w,w> \geq c \|w\|^2 \quad \text{for some} \quad c > 0 .$$

Then ϕ_0 is linearly and nonlinearly stable.

Proof: First of all we note that (3.1) implies the energetic stability. To prove the linear stability we use the linearized energy $<E''(\phi_0)w,w>$ as a Liapunov functional. The linearized energy is conserved for (H_{lin}) and the norm can be controlled by estimate (3.1). Each solution of the linearized system remains bounded which implies the linear stability of u_0 .

To prove nonlinear stability we observe that (3.1) implies the existence of $\varepsilon > 0$ such that

$$(3.2) \qquad E(u) - E(\phi_0) \geq c \|u - \phi_0\|^2$$

for all u with $\|u - \phi_0\| < \varepsilon$. Estimate (3.2) will then imply the nonlinear stability. □

If there is an additional conserved quantity Q , e.g. the charge, then an estimate of the type (3.1) for all perturbations which keep the quantity Q fixed will be sufficient to prove nonlinear stability.

To illustrate this fact let us consider the following dimensional Klein Gordon equation

$$(3.3) \qquad U_{tt} - U_{xx} + U - |U|^{2p} U = 0 \qquad \text{on } \mathbb{R} .$$

It can be easily checked that for each ω with $\omega^2 < 1$ equation (3.3) possesses a unique solution (up to translations in space and phase) of the form $e^{i\omega t} u_\omega(x)$ with u_ω symmetric, positive function strictly decreasing on $\mathbb{R}^+$. One can prove that if

(3.4) $d(\omega) \equiv \int u_\omega'^2(x) \, dx$

satisfies $d''(\omega) > 0$ then u_ω is energetically and nonlinearly stable.

The energy and the charge associated to (3.3) are given by

$$E(u,u_t) = \frac{1}{2} \int |u_t|^2 + |u_x|^2 - \frac{1}{p+1} |u|^{2p+2} \, dx$$

$$Q(u,u_t) = \text{Im} \int \bar{u} \, u_t \, dx \quad .$$

Indeed in [BSV1] we show that estimate (3.1) is valid for all tangent vectors of curves $(u_1(\lambda),u_2(\lambda))$ which keep the charge fixed and are orthogonal to the orbit of $u_\omega(x)$

$$\frac{d^2}{d\lambda^2}\bigg|_{\lambda=0} E(u_1(\lambda),u_2(\lambda)) \geq C(\|y_1\|^2 + \|y_2\|^2)$$

where $y_j = \frac{d}{d\lambda} u_j(x) \Big|_{\lambda=0}$.

Results in the same direction have been obtained for a lot of nonlinear field equations [Be, Bo, GSS, W2] .

Instability results for nonlinear Klein Gordon equation have been proved if $d''(\omega) < 0$ in [SS] .

For nonlinear Dirac equations the situation is much more difficult. Solitary waves of the Gross-Neve model or the Thirring model are expected to be nonlinearly stable but as we showed in [BSV2] they are always energetically unstable which can be traced back to the indefinite 'Kinetic' term in the energy.

4. Soliton stability

We finish our lecture with a few remarks on soliton stability. As pointed out in the introduction the soliton stability is related to the complete integrability of the corresponding wave equation. In such case when the solitary waves interact, they are always scattered elastically, preserving asymptotically their shape. The integrability also permits an analytical study of the multisoliton interactions. [DEGM]

In many real physical systems the basic models are not integrable, and in other cases the integrability condition (and then the solitonic stability) is destroyed due to physical perturbations (impurities, external fields, ...). In certain cases of small perturbations, the solitary wave interactions can be treated analytically, but, in general, we must study numerically the dynamics of the solitary waves collisions. The understanding of the interaction mechanisms can be used to given a measure of how far is the system from an integrable one. The natural mathematical

frame of the collision phenomenology would be a KAM theorem for infinite dimensional systems.

There are many numerical studies, begun in the mid 1970's, about the collisions
of solitary waves in non-integrable systems [AKL,M,CP,CPS] . Here we summarize
some of the obtained results:

(1) Collisions of solitary waves, in one spatial dimension, in confining models
    [ST,AKL] : In such collisions new localized objects are generated, which are
    pulsating in time, and they appear to be stable.

(2) Collisions of the solitary waves associated to the one-dimensional Dirac field
    with a scalar self-interaction [AC] :

$$i \; \gamma^{\mu} \; \phi_{\mu} \; \psi - m \; \psi + 2 \; \lambda \, (\overline{\psi} \; \psi) \; \psi \; = \; 0 \; .$$

The numerical experiments showed inelastic interactions and bound state produc-
tion in binary collisions. Also it was observed charge and energy interchange
except for some particular initial velocities of the solitary waves.

(3) Kink – Antikink $(K \; \overline{K})$ collisions in the one-dimensional nonlinear Klein-Gordon
    equations

$$\phi_{tt} - \phi_{xx} - \phi + \phi^3 \; = 0$$

$$\phi_{tt} - \phi_{xx} + \sin \phi + 2 \lambda \; \sin \frac{\phi}{2} = 0 \; .$$

The following features are observed in the center mass frame (in which a Kink
with velocity V collides with an antikink with velocity - V)

(a) For lower velocities than a critical value (V_c) there is a "trapping" and
    an oscillatory state is created.

(b) Inelastic interactions: the final velocity of the solitary waves is less then
    their initial velocity.

(c) For well-defined velocities, below V_c , the $K\overline{K}$ reflect once, escape to fini-
    te separation, and finally reflect again before separating to infinity.

References

[AKL]   M.J. Ablowitz, M.D. Kruskal and D.F. Ladik, SIAM J. Appl. Math. 36, 428
          (1979)

[AC]   A. Alvarez and B. Carreras, Phys. Lett. 86A, 327 (1981)

[Be]   T.B. Benjamin, Proc. Roy. Soc. London A328, 153 (1972)

[BSV1]   Ph. Blanchard, J. Stubbe and L. Vazquez, Journ. Phys. A21, 1137 (1988)

[BSV2]   Ph. Blanchard, J. Stubbe and L. Vazquez, Phys. Rev. D36, 2422 (1987)

[Bo]   J.L. Bona, Proc. Roy. Soc. London A344, 363 (1975)

[CP]   D.K. Campbell and M. Peyrard, Physica 18D, 47 (1986)

[CPS]   D.K. Campbell, M. Peyrard and P. Sadano, Physica 19D, 165 (1986)

[DEGM]   R.K. Dodd, J.C. Eilbeck, J.D. Gibbon and H.C. Morris, "Solitons and Non-
          linear Wave Equations", Academic Press, London, 1982

[GSS]   M. Grillakis, J. Shatah and W. Strauss, J. Funct. Anal. 74, 160 (1987)

[J]   R. Jackiw, Rev. Mod. Phys. Vol. 49, 681 (1977)

[JR]   R. Jackiw and P. Rossi, Phys. Rev. D21, 426 (1980)

[M]   V.G. Makhankov, Phys. Rep. 35, 1 (1978)

[ML]   D.W. McLaughlin and A.C. Scott, Phys. Rev. A18, 1652 (1978)

[ST]   Y.A. Simonov and J.A. Tjon, Phys. Lett. 85B, 380 (1979)

[SV]   J. Stubbe and L. Vazquez, to appear in "Mathematics + Physics" Vol. 3,
          L. Streit, editor, World Scientific, Singapore

[W1]   M.I. Weinstein, SIAM J. Math. Anal. 16, 472 (1985)

[W2]   M.I. Weinstein, Comm. Pure Appl. Math. 39, 51 (1986)

Lecture Notes in Mathematics

Lecture Notes in Physics

H. L. Cycon, R. G. Froese, W. Kirsch, B. Simon

Schrödinger Operators

with Applications to Quantum Mechanics and Global Geometry

Springer Study Edition

1987. IX, 319 pp. 2 figs. Softcover DM 56,–
ISBN 3-540-16758-7

Contents: Self-Adjointness. – L^p-Properties of Eigenfunctions, and All That. – Geometric Methods for Bound States. – Local Commutator Estimates. – Phase Space Analysis of Scattering. – Magnetic Fields. – Electric Fields. – Complex Scaling. – Random Jacobi Matrices. – Almost Periodic Jacobi Matrices. – Witten's Proof of the Morse Inequalities. – Patodi's Proof of the Gauss–Bonnet-Chern Theorem and Superproofs of Index Theorems. – Bibliography. – Subject Index. – List of Symbols.

S. A. Albeverio, F. Gesztesy, R. Høegh-Krohn, H. Holden

Solvable Models in Quantum Mechanics

1988. XIV, 452 pp. 51 figs. (Texts and Monographs in Physics) Hardcover DM 158,– ISBN 3-540-17841-4

Contents: Introduction. – The One-Center Point Interactions. – Point Interactions with a Finite Number of Centers. – Point Interactions with Infinitely Many Centers. – Appendices. – References. – Index.

Springer-Verlag
Berlin Heidelberg New York
London Paris Tokyo Hong Kong

Springer